The Honey Factory

The Honey Factory

Inside the Ingenious World of Bees

JÜRGEN TAUTZ & DIEDRICH STEEN

Published by Black Inc.,
an imprint of Schwartz Publishing Pty Ltd
Level 1, 221 Drummond Street
Carlton VIC 3053, Australia
enquiries@blackincbooks.com
www.blackincbooks.com

Copyright © Jürgen Tautz and Diedrich Steen 2017
First edition published by Gütersloher Verlagshaus, a division
of Verlagsgruppe Random House GmbH, München, Germany, as
Die Honigfabrik. Die Wunderwelt der Bienen – eine Betriebsbesichtigung
This edition published in 2018
English translation by Dr David C. Sandeman, 2018
Jürgen Tautz and Diedrich Steen assert their right to be known
as the authors of this work.

ALL RIGHTS RESERVED.

No part of this publication may be reproduced, stored in a retrieval system, or
transmitted in any form by any means electronic, mechanical, photocopying,
recording or otherwise without the prior consent of the publishers.

9781760640408 (paperback)
9781760640903 (hardback)
9781743820575 (ebook)

Cover illustration by Alice Oehr
Cover design, text design and typesetting by Tristan Main
Author photo courtesy of Random House, Germany

Printed by Sheridan in the United States of America

Contents

Introduction 1

1 **The Factory and Equipment of a Bee Colony** 7
The evolution of beekeeping 7
Comb network technology 17

2 **Teamwork in the Honey Factory** 29
Female power in the bee colony 30
Cuddling up together 32
Climate control in the nursery 41
Honey bees and sleep: Too tired to attack 62
The waggle dance: Rethinking old concepts 67
The drones: Callboys for the queen 86
Fatherless companions 90
The queen bee: A monarch with limited power 97
The 'Bien': An intelligent superorganism 104

3 The Honey Factory Production Line   119
What comes out of a honey bee   121
Collected resources: Propolis, pollen and bee bread   127
The crème de la crème: Honey   133

4 Founding a Daughter Company
The Swarm   157
Driven by instinct   158
Swarm intelligence: How bees move house   161
Events in the old factory: The terror of succession   169
Killjoy beekeepers   172

5 Bees as Aggressors   183
Hostile takeovers: How bees steal from one another   185
The bees and the beast   186

6 **The Struggle for Survival** **197**
  The myth of the dying honey bees 199
  Extinction remains a possibility 205
  Back to the future: Rediscovering old practices 216

Epilogue
Honey Bees: A Way of Life **229**

Acknowledgements 239
Bibliography 241
Endnotes 253
Index 263

To my children, Mona, Silke and Meiko, and my grandchildren, Anton and Oscar

Jürgen Tautz

For Dirk Steen on his eightieth birthday

Diedrich Steen

Introduction

It could be a perfect evening. After work, I stroll a few paces behind the house to the hives under the oak trees to watch the last returning nectar-laden foragers tumble down onto the landing stage. They get a short greeting from the guard bee and then head into the dark hive, where their sisters wait to accept their precious load.

And the perfume! It's the end of April, the cherry trees are still blooming, and the dandelions and apple, plum and pear trees are all about to release or already releasing their nectar. On good foraging days, the scent near the beehives is like the smell of candy floss at a spring fair. It's so good to breathe in the warm, sweet and heavy aroma of a nectar-rich day and to hear the low hum coming from the hive. All is well.

Except all is not well and I worry. First, we didn't get a real winter this year. At Christmas, daytime temperatures were in the double figures and there was no frost at night. The queens did not stop laying. Emerging larvae had to be fed. The number of bees in the colony stayed high and winter food stores were quickly being depleted. Would they last until the meadows bloomed in March? By February, several beekeeper friends had already reported colonies starving. Did I feed them too little in late summer? There is no more shameful sight for a beekeeper than a starving colony.

At the beginning of April there were a few warm days. The willows bloomed, the food situation improved with some nectar flowing into the hive, and the queen was now laying up to 1200 eggs per day. She would increase this to a total of 2000 a day. But the weather needed to remain good! And then, there was an Arctic cold front for ten days. Would the many larvae now in the colony make it through? Could the nurse bees feed them enough and also keep the nest warm? The nights are frosty again. Will the food store last?

Finally, on the first day of May, there is relief. The forecast is that a high-pressure front will bring a change in the weather. It is now ten o'clock on a Sunday morning, the sky is blue, the thermometer stands at 12°C and in two more hours it will reach 15°C. There will be no stopping the nectar gatherers. It can all begin.

*

INTRODUCTION

This, dear readers, is how beekeepers thought and felt in many places across Germany early in 2016. Yes, *felt*, because the connection keepers feel with honey bees is a passionate and emotional one – and not only because of all the current media reports about bees dying and the unavoidable approaching demise of the human race. Those who begin to keep bees and still have colonies after the dramas of the first three years do not possess them anymore. Instead, they are possessed by the bees! There are beekeepers who have reached 100 years of age and still tend their colonies, perhaps with a cross-country walking aid and an elderly son as a slave to lift the heavy hives.

Beekeeping is fascinating because bees continue to surprise even those who have kept them for decades. When describing their experiences, amazed beekeepers will often say, 'They have never done that before.' Every bee colony has its own character and every year follows its own rhythm. Bees are never boring. A bee colony is a complex organism, rather like a book that one can read again and again each year and find new and interesting stories at each reading.

We, Diedrich Steen and Jürgen Tautz, would like to take you with us into the world of these stories. Diedrich is a publishing director and has been a beekeeper for twenty years; Jürgen has been a professor at the University of Würzburg for twenty-seven years, has a PhD in biology and is an internationally renowned bee researcher. Diedrich will tell you what he says when asked, 'Hey, I've heard you keep bees. Is it true that...?' Jürgen will ensure what Diedrich says is really true. Primarily, though, we will juxtapose a beekeeper's practical knowledge against the exciting background of science and bee research. You can distinguish these two accounts in the text by the use of different typefaces.

Together we invite you to visit a bee colony – a honey factory – and take a guided tour through its works. We will see the factory workshops and equipment and get to know the staff, directors and products. We will learn who does (or does not) work with whom; hear about slackers and freeloaders, but also about zealous specialists; and visit a world filled with amazing checks and balances. Although at first sight life in a bee colony seems to be anarchy, in fact bees know what they are doing. They execute their plans with surprising cleverness, fascinating ability and impressive teamwork.

We chose to use 'honey factory' as a model because from a beekeeper's perspective that is exactly what bee colonies are: businesses with up to 50,000 female workers in honey production and a few male seasonal helpers. The bees most likely have a different view. If we were to ask one of them if they worked in a honey factory, she would probably just wave her feelers and not understand the question, because the production of honey is not the reason

INTRODUCTION

for the bee's existence. They are concerned with ensuring the survival and proliferation of the colonies; honey is merely the means – the source of energy and fuel to achieve this. Apiarists are the ones to make honey the purpose in life and the ones who exploit the bees as factory workers to this end.

Is it unfair of them to do so? A reprehensible interference in the natural affairs of a living organism – a creature that currently receives much sympathy? Some see it that way. However, what is true for most animals that find themselves in the care of humans is also true for bees. We have turned them into economically useful animals. Bees let themselves be used, but do they allow themselves to be *exploited*? Or is it instead the beekeepers who adapt themselves to the bees and their needs if the honey factory is to be successful. We shall see ...

A few words about the organisation and intent of this book. We will describe what goes on in a bee colony and try to convey an impression of its overall coherence and integration. Our book is therefore not an apiarist's manual from which one can learn how to keep bees. Naturally we hope that beginners and perhaps even some experienced beekeepers will find something here that is stimulating and helpful for the pursuit of their hobby. But above all this book should communicate an understanding of honey bees to all those who are interested in these amazing insects and for whom honey tastes so good.

To arrive at such an understanding, the chapters should be read in order because they build on each other. The chapter on swarming, for example, is more meaningful if you know what the combs are for and how bees communicate with one another. However, those who wish to jump about will find an index at the end of the book to help guide them.

And now, welcome to the honey factory and the world and works of honey bees.

1

The Factory and Equipment of a Bee Colony

The evolution of beekeeping

Plundered caves

In their natural state, bees – at least the European ones – live in caves or tree hollows. During our evolution, humans also emerged from caves – as so-called intelligent anthropoids, to develop into more or less functioning social beings. But bees developed into social beings in crevices in cliff faces, under stones or in hollow trees long before humans achieved their upright stance. A bee that flew from flower to flower 45 million years ago has been found in fossil beds from the Eocene epoch. From this, one can assume that the ancestors of bees we know today may well have stung the dinosaurs that trampled their homes.

About 1.7 million years ago, the first members of the *Homo sapiens* race appearing on the scene in Europe would have quickly realised that bees hoarded a precious treasure in their nests. Nowhere else in the human world at that time was such a calorie-rich and delicious food to be found.

Perhaps humans learnt how to get to the honey by watching bears – rip open the hive, grab the combs and run away before being stung – for bees of that time were already equipped with stings and surely used them to defend their homes. But those who sought honey also needed to possess endurance and be prepared to take risks. Stone Age drawings in the so-called 'Spider Caves' near the village of Bicorp in Spain show a honey collector climbing down a sort of rope ladder to a bee colony. Today, a similar practice is still carried out in the Nilgiri Biosphere Reserve in southern India. Asian bee colonies, in contrast to European bees, each build a single freely suspended comb beneath a projecting rock shelf. Honey collectors of the Kattunayakan, a south Indian indigenous people, harvest honey by climbing down a bamboo rope and breaking off the combs with a hooked stick.[1] The practice is hazardous for the collectors and destroys the colony's comb and probably the colony itself – at least in those parts of the world where the vegetation is deciduous. In such climates, bees need honey as their food reserves for winter. It is honey that enables bees to generate warmth and keep themselves alive during the cold months of the year. When our Stone Age ancestors robbed bees of their honeycombs, in most cases it would have been the end of the colony. A plundered colony with extensively damaged combs cannot survive.

First beekeeping practices

It did not take humans long to learn that a regular supply of honey cannot come just from stealing it. One has to offer something in return. So people started developing 'caves' which bees could live in that were made by hand from pottery, tree bark or baskets smeared with clay. Although not apiary in the modern sense, this was nevertheless a beekeeping strategy. People no longer simply sought out the homes of the bees and stole their honey. Instead, they enticed the bees to a specific place where a number of clay pipes were suspended vertically, close together, in trees. Swarming bees looking for places to settle entered the clay pipes and occupied them. The 'beekeeper' thus gained a new colony. With enough colonies in pipes, the keeper could refrain from harvesting honey from some of them, allowing those colonies to survive the winter. In the following year, these colonies would swarm early and so contribute to a rapid increase in the number of the beekeeper's colonies. The keeper could again harvest all the honey from some colonies in late summer and allow others to survive. The cycle would then begin again.

Skeps, or basket apiaries, were a similar beekeeping strategy and were widely used across Europe from medieval times to the end of the nineteenth century. Basket beekeepers were not only interested in honey but also in wax, which was much in demand for the production of candles for churches and cloisters. A few skep apiaries still persist today in the Lüneberg Heath in north Germany. Basket beekeepers begin in spring with a few colonies. The number of bees in these colonies increases rapidly in February, the baskets become crowded and the pressure to swarm grows. Should a swarm leave a hive, the beekeeper catches and accommodates it in an empty basket, which the bees usually accept and occupy. The beekeeper has a new colony. This process repeats itself several times from the end of April until about the middle of July. In late summer when the heather blooms, beekeepers have many more colonies than they had in spring. These are now arranged along horizontal boards, protected with a roof and set out in the heather, where the bees collect nectar for heather honey, to be harvested when the heather no longer blooms.

In the past, harvesting honey was often fatal for bees. Apiarists first dug a shallow pit and burned strips of paper in it that had been soaked in sulphur. The basket hive was then placed over the pit in the rising sulphur fumes. The bees promptly suffocated and the combs could then be broken out. But it must be said that more often the beekeeper would set the honey and bee-filled basket over a second, upside-down empty basket and then thump them both onto the ground. Most of the bees fell out into the empty basket and could be added to colonies chosen to survive the winter. Those that remained in the old hive died in the sulphur fumes.

This method of beekeeping leads to the destruction of both the combs and the colony. This was not seen to be a problem while beekeepers wished to harvest wax. But beeswax as the raw material for candles became less desirable from the middle of the nineteenth century, when the French chemist Michel Eugène Chevreul found out how to prepare fatty acids from animal fat and soon stearin, the material from which most candles are made today, was discovered. The development of paraffin wax followed and, finally, at the end of the nineteenth century, electricity and light bulbs brightened houses at night. Honey also lost its economic importance when, in 1801, Franz Carl Achard founded the first sugar factory in the world to refine sugar from beets. Until then, those wishing to sweeten their meals had to make do with expensive imported cane sugar or relatively scarce and also expensive honey. Sugar refined from beets offered a much cheaper alternative. Where sweetness had been a luxury enjoyed regularly only in the homes of the well-off and of beekeepers, it was now a widely available consumer product.

Apiarists had to adapt to this new situation. To avoid throwing the wax away, they needed to find a way to harvest honey without destroying the colony's combs. Beekeepers seeking to compensate for the fall in the price for honey needed to increase production in order to have more to sell.

It would take too long to describe the developments that followed somewhat chaotically in the nineteenth century, when the scientific observation of natural phenomena became a serious occupation. Systematic study of the activity of bees in a colony was part of this effort. Countless clubs and societies were founded that busied themselves with bee research and the improvement of beekeeping practices. Much that was regarded to be scientifically established at the time turned out to be humbug – and inventions claiming to revolutionise beekeeping vanished quickly from the scene. Two new developments persisted, however, significantly influencing modern beekeeping and leading to establishing colonies in stacked boxes, and to installing combs in removable frames.

Hives gain space and frames

Basket beekeepers noticed that bees would accept help and guidance when building their combs. Wooden rods pushed through the walls at the top of the baskets so that the ends stuck out on either side provided the interior of the hive with a series of horizontal beams. The bees began to build their combs along the rods and continued vertically downwards, ending up with completed combs hanging down from the rods. If the beekeeper placed the rods parallel to one another, the combs would hang straight and separately down, without bridges between them. It was now possible to break single combs out of the hives without damaging the others.

Inventive beekeepers had already started to develop rectangular, stackable baskets. Beehives became multistorey, with the floors separated from one another by a board. A hole in each board allowed the bees to slip through and move between the floors. An entrance was placed at the bottom of the stack, through which the bees could enter and leave the hive.

Bees primarily need their combs for two purposes. First, the queen must have somewhere to lay the eggs that will develop into young bees, and she lays these in comb cells. Second, food reserves, pollen and honey are stored in the comb cells. We will learn more about this later. The contents of the combs are organised by the bees as follows: Cells accommodating eggs and brood are located in an area in the centre of the comb. Above this is a narrow pollen crescent, as apiarists call it, of cells containing pollen. Above this again is the honey crescent: cells filled with honey. The bees arrange the contents of the cells so that they have the food where it is needed, right next to the brood that have to be fed. As time passes and more honey is brought into the hive, the brood nest and the adjacent cells migrate downwards and the honey crescent expands above. The consequence of this for a beekeeper with a multistorey hive is that they can begin with a single storey and then add another storey to accommodate the growing brood nest. When so much honey has been collected that combs in the upper floor are full and contain only honey, these can be removed without disturbing the brood nest and throwing the colony into panic.

The basket beekeepers' rod technique was soon creatively combined with the multistorey hive. Flat wooden boxes were constructed with a grid of rods supported on their upper edges. The boxes were stacked on a base board with an entrance hole. Whenever the upper box (the super) was filled with honey, it could be removed, the honey harvested and the box returned, empty.

The end result was a stacked wooden hive with multiple storeys or 'supers' and horizontal guides for comb building. Beekeepers could now operate more considerately than with basket hives because harvesting honey no longer disturbed the entire colony. One problem remained: the rods bearing the combs still had to be cut out of the supers to harvest the honey.

In the middle of the nineteenth century, the Polish priest Johann Dzierzon and Baron August von Berlepsch in Thuringia found the solution to this difficulty. Instead of laying a grid of rods on his supers, Dzierzon used individual, separated rods. Now he could cut the combs from the walls of the super and take out the combs separately. Berlepsch wanted to avoid the cutting-out procedure entirely. He made a frame by extending the rods down on each side and joining these laterally at the bottom. He could now hang these frames in the super and let bees build their combs in them, which, most of the time, they did without joining them to the sides of the super. Thus, removable combs were invented.

Modern honey factories

Multiple storeys and removable frames – these two inventions laid the foundations of beekeeping as it is practised today. The aim was then to improve on these inventions. How large should hives be to ensure optimal colony development? What is the size of an ideal frame? How should the proportions and construction of the hive and frames be determined to provide an appropriate living space for bees but at the same time be easily and economically operated by beekeepers?

A period of wild experimentation by individual beekeepers began that basically has never ended and is frequently characterised by unkind abuse of anyone who approaches things a different way. Presently there are eighty different sizes of comb frames around the world. Each fits only one size of super, and all apiarists are convinced that *their* system and practice is the only soul-satisfying way to keep bees. The arguments continue.

However, beekeeping with magazine hives has been adopted internationally. Magazine hives consist of a baseboard with an entry. A rectangular case with a removable lid is set on top of the baseboard. The case is a four-sided box, open top and bottom, in which the comb frames can be suspended. The cases can be constructed from wood or plastic. Two sizes of frames have been adopted in Europe: the German normal size, which measures 37 cm long by 23.3 cm wide; and the Zander, which is 42 cm long by 22 cm wide. Beekeepers using the normal-sized frames usually fit eleven frames into each case; those that work with Zander frames, nine. The normal size provides the bees with a greater area, but also more frames for the beekeeper to handle.

Magazine hives are built for the production of honey and the ingenious aspect of their construction is that they can be expanded. A colony that has lived through winter in a single case can be provided with a second case with frames, or super, as it grows in spring. This is simply placed on top of the first. When the flowers bloom and the bees bring in nectar, a second super can be added. If things go really well and the beekeeper is tall enough, a third can be added to the stack. With four supers in June, up to 40,000 bees can be active in the colony.

Such colonies are huge and the comb areas enormous, but nothing is inaccessible to the beekeeper, because each super can be taken off the stack individually and every single frame is available for inspection. Removing the lid of the hive exposes the frames in the uppermost super. Leaving the lid on but lifting up the top super exposes the frames in the super below it, and so on down through the hive. Frames can be pulled up and out separately from any of the supers and examined.

And so it was that the interest and inventiveness of humans led bee colonies from caves and hollow trees to honey factories in magazine hives with supers. Whereas the initial relationship between people and bees was characterised by destructive plundering of their hives, magazine-hive beekeeping has facilitated the ongoing care, and use, of bees. Magazine hives were the key, because they allowed access to the comb, the most important component of the honey factory, without destroying the colony.

Comb network technology

Bees build their factory equipment with sweat and hard labour. From the age of about ten days, a young worker bee is able to 'sweat' tiny wax plates from eight small groups of glands on the underside of her abdomen. The wax glands atrophy as the worker ages and applies herself to other tasks in the hive, but they can be reactivated if needed. During swarming, for example, when new combs have to be built, older bees redevelop their wax glands and are able to join the building gangs. We will describe this in more detail later.

The small extruded wax plates are the raw material for the comb and consist of more than 300 chemical components. Almost all are hydrocarbon compounds[2] and some of these molecules are light and evaporate easily. They are responsible for the pleasant aroma of the wax. Most of the wax molecules are long-chain hydrocarbons, some containing up to fifty-four carbon atoms. The physical properties of beeswax are determined to a great extent by the spatial organisation of these large and long molecules. When they all lie parallel to one another, the wax exhibits an almost crystalline nature, and is structurally more stable.

The molecules in wax the bees extrude are not geometrically oriented with one another. Such wax is unsuitable for building and must first be processed by bees: they take the plate wax from their colleagues in their mouthparts, chew it and add an enzyme that splits the long-chain molecules. The wax becomes more pliable by warming and kneading it. Bees at the building site also ensure that the temperature lies between 30°C and 40°C. Warmth softens the wax, making it easier to work. Here, though, a problem arises for the bees.

A compound building material

Wax is a dynamic material. Its properties change when the lighter components evaporate, and it changes its consistency with fluctuations in the ambient temperature. When cold, it is hard and brittle; if it's too warm, it liquefies. Combs that are too soft cannot fulfil their primary function for the colony as storage cells because they are unable to support the weight of the honey stored in them. However, bees have a way of increasing the load-bearing capacity of the comb wax. They use a type of 'prestressed' wax made from wax and propolis. Propolis, which will be described more fully in Chapter 3, is a material bees derive from plant resins and is mixed with the wax to produce a compound building material. The amount of propolis used and its exact distribution in the wax is determined by where it will be used. More propolis is added to the wax in warmer climates.

Bees reinforce the upper rims of the comb cells with a thin layer of propolis, creating a coherent and stabilising net of small bridges along the edges of the comb cells. The effect of the small bridges on the properties of the comb can be estimated by comparing their stiffness with that of artificial bridges moulded from melted bridge material. The bridges the bees build are significantly more difficult to bend than the copies made by researchers from the same material. The bridge stability is clearly not solely dependent on the material, but on how they are constructed. Bee-built bridges have an internal structure that we are unable to reproduce.

A second strategy bees employ to stabilise their comb structures is very well known to us. Propolis is not only incorporated into the rims of cells but also into the cell walls in the same manner that engineers manufacture prestressed concrete. The danger of cracking following extreme fluctuations in temperature can be significantly reduced in concrete by including short steel rods before it has set. The rods become randomly distributed throughout the mixture and stabilise the set concrete to prevent fracturing when under tension. Bees also exploit this technique. The wax can be regarded as the equivalent of the concrete, and the tiny propolis particles the steel rods. The spatial distribution of propolis particles in the paper-thin walls of comb cells remain as separate entities, just like metal rods in concrete.[3] The combination of propolis and wax provides combs with a load-bearing capacity more or less equivalent to a yoghurt container. Two kilograms of honey can be safely stored in a comb built from just 30 grams of wax.

The comb as a nursery

The comb is not only for storing honey; it is also the colony's nursery. The queen deposits a single egg at the bottom of each

THE FACTORY AND EQUIPMENT OF A BEE COLONY

cell, and three days later this hatches into a larva, which is cared for and fed by nurse bees for another six days. The nurse bees then cap the cell with wax and the larva pupates. A new worker emerges from the cell twenty-one days after the egg was laid. Brood combs contain cells for both worker bees and drones, which are male bees. Drones are somewhat larger than workers and develop in larger cells. More about them later. Bee combs constructed under natural conditions contain many small worker brood cells and a few areas with larger brood cells for drones. For reasons that will become apparent later, beekeepers prefer not to have both worker and drone cells in the same comb. To avoid this, they use the frame as a guide and to influence the size of the cells the bees will build. A colony that is expanded by adding a new super does not get one with empty frames. Instead, each frame is supplied with a vertical central partition that fits exactly into it, dividing it into two. The partitions are supported on thin wires stretched across the frame, and when these are electrically heated they melt into the wax. When cooled, the wires prevent the partitions from falling out of the frames. The secret here is that the partitions are embossed on each side with a pattern that has the dimensions of worker bee comb cell bases. Bees accept these centrally located foundations and build only worker bee cells on both sides. The result is a perfectly regular comb and, because the bees can simultaneously build cells on both sides of the foundation and across the entire area of the frame, the new comb is rapidly completed, and stable due to the supporting wires.

Interference on the house telephone

As well as being a place for storage and a nursery, combs also serve as a communication network.[4] Apart from a little light that enters through the small entrance in the baseboard, the honey factory is dark inside and dominated by a teeming crowd moving about in complete darkness. How can they communicate without seeing or hearing?

One possibility is the exchange of chemical signals. Specific odours representing particular signals are either there or not. And, in fact, bees do communicate with odours: we will return to this later. However, although bees cannot hear, they can still feel.

Bees have highly sensitive tactile organs on their feet and legs, with which they can detect different frequencies of comb vibration.[5] What we perceive as sound, bees detect as vibration. A newly emerged queen, for example, emits a tooting noise that is loud enough to be heard by beekeepers, but felt as a comb vibration by bees. With this signal, the new princess announces 'I am here' to her colony.

The comb is thus *the* communication platform in a beehive and the propolis-reinforced cell rims we described previously constitute the 'landline' of the house telephone within the bee colony.[6] How it functions will become clearer later, when we examine the waggle dance in detail.

Bees dance in order to transmit information, and each dancing bee attempts to recruit followers. The more bees that dance, the more rapidly the information is spread through the colony. The way dances attract attention is particularly sophisticated. Briefly, the waggle dance consists of a running phase and a waggle, or standing, phase. During the standing phase, the dancer stands virtually still, lifting first one and then another of her six legs in an uncoordinated way, as though searching for a better foothold on the comb. She moves slightly and slowly forward and throws her body from side to side up to fifteen times a second.[7]

The dancing bee thus maintains firm contact with the comb while pulling herself towards cell rims alternately to the left and right of her. The rims are brought under tension, somewhat like stretching a rubber band. Bees do not weigh much but are able to exert a force of 4 millinewtons, equivalent to the weight of five bees, against the cell rims, displacing them by about two-thousandths of a millimetre.[8] The signal the dancer produces to attract followers is a short, pulsed vibration coming from the flight muscles, with a fundamental frequency of about 250 hertz.[9] The dancer emits these pulses to coincide with the moment she pulls most strongly on a cell wall; that is, when the force coupling is maximal and the vibration pulses optimally transmitted to the telephone network.

The following analogy summarises the process: suppose we tune a guitar string to the tone 'C' and a singer also sings the note 'C'. The guitar string, without us touching it, will vibrate, activated by the body of the instrument, which responds to the sound from the singer. The bee dancer employs a similar strategy. She places the cell wall under tension, like a guitar string, and then 'sings' a 250-hertz vibration, the optimal frequency for the system, into the telephone system.

The production of the vibration is accompanied by unavoidable wing movement and air currents that can be measured as sound. But the dancers in the dark hive attract their followers solely by the vibration of the substrate they stand on and not by airborne sound or some other physical phenomenon. This is clear from behavioural observations of a dancer and follower who stand on the same dance floor, compared with a pair the same distance apart but not standing on the same floor. Followers are attracted to the dancer only in the first case and exhibit the following typical behaviour (Figure 1). Bees standing close by but on a neighbouring comb and to which the dancer turns her back respond only if they happen to touch her with their antennae. They are not standing on the same comb,

THE FACTORY AND EQUIPMENT OF A BEE COLONY

receive no signals over the house telephone and need direct contact with the dancer to get her message.

Figure 1

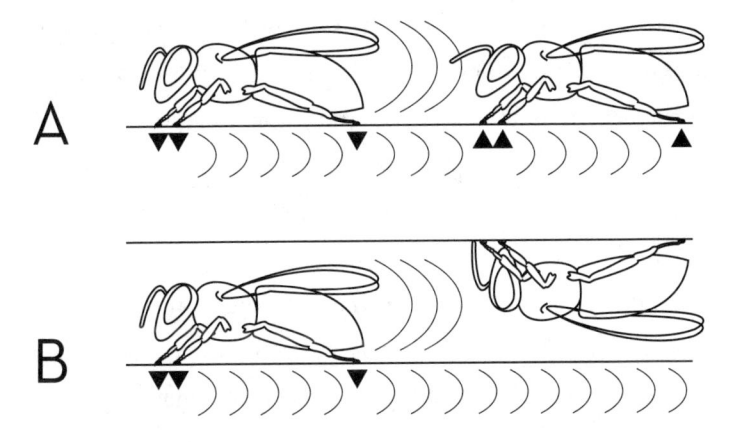

A.

B.

A. Acoustic stimuli through the air and vibration of the substrate, in this case the comb, from a dancer (left) reach and attract a dance follower (right) when both are standing on the same substrate.

B. Dance followers (right) not sharing the same substrate with the dancer (left) receive only the acoustic signals and are not attracted to the dancer.

The house telephone works best when dances are held on open, uncapped comb cells. The telephone net is impaired on other dance floors because they cannot oscillate freely. Dances on capped cells, on the wooden frames or, as we will see later, on the bodies of colleagues in a swarm attract significantly less attention.

Matching the vibrational message of the dancer to the communication network depends on the impedance of the cell rims. Impedance means resistance. The smaller the force required to trigger a specific motion of a structure, the lower the impedance of that structure. The impedance and the conduction capacity of the cell rim network depend on the temperature of the wax, the cell size and the frequency of the vibrations that spread through the net. Frequency dependence dominates. The impedance of the network is lowest to frequencies around 250 hertz. The strength of the signal and the conduction capacity of the cell rim network are in an optimal relationship with one another in this frequency range. Somewhat surprisingly, whether dances take place on small worker bee combs or larger drone cells has little influence on the conduction capacity of the comb. Temperature exerts a markedly greater effect. The impedance of the rims increases if the wax is cold and vibrations are less well conducted. In winter, when it can freeze in honey factories, the telephone is largely inoperative. On the other hand, the warmer the wax, the weaker the tension dancers can induce in the cell walls. It is precisely for this reason that reinforcing the cell rims with propolis is of such importance for bees.

Filling the cells with honey has, unexpectedly, no influence on the transmission of the vibrational signals. From a purely physical point of view the cell rim network behaves as though it were able to swing freely and independently of the cell walls. There is also no negative effect on the communication network if beekeepers provide comb foundations of stiff plastic for the bees to build on.

The transmission of signals is impaired in only two situations. First, if cells are capped, the walls become immovable and the telephone is deactivated. The same applies when a comb is completely fused to the frame on all sides. Beekeepers repeatedly see how bees cope with this. Gaps and holes between the comb and the frames are gnawed out by bees to provide the net with the necessary flexibility. Telephone-repair bees have ensured that the dance floors keep swinging and the house telephone remains intact.

2

Teamwork in the Honey Factory

Honey factories are seasonal concerns. In winter, most activity stops and the personnel are primarily occupied with keeping each other warm. New life creeps slowly in as the days become longer. After the winter break, the queen begins to lay again and soon the first young bees emerge from their pupae. The colony grows as its members increase, and reaches a peak as the flowering season comes to an end. The colony's population declines in late summer and early autumn, until only the winter bees remain.

The first section of this chapter is concerned with these winter worker bees.

Female power in the bee colony

One day in late September or early August, the time comes: a winter worker bee begins her life. Three weeks after the queen laid the egg which led to her existence, the winter worker bee gnaws her way through the protective wax cap that her older sisters had fastened over her pupal cell. Once the hole in the cap is large enough, she squeezes through and pulls herself out of the cramped cell. Somewhat uncertain on her feet, and with creased wings, she stands on the comb surrounded by a tumult of sisters, who pay scant attention to the new arrival. She is just one of many who have emerged on this day, as on every other day for the last six months.

Winter bees

But our small bee and the others leaving their brood cells at this time are special. They are the winter bees. Unlike their sisters who hatched in spring or summer, they will not live for only six to eight weeks; instead, their life expectancy is six or seven months. Not until March or April in the following year will they, one after the other, fail to return to the hive after a foraging flight. Before they come to their end, they have a special task to perform at a particular time. It is now autumn, very few flowers are still blooming outside and the queen lays fewer eggs than in spring and summer. The colony population decreases and prepares itself for winter. Foragers bring in just enough nectar and pollen to nourish the hive bees and the growing brood. The winter supplies – 20 kilograms of sugar syrup that the beekeeper has given the colony in exchange for their honey – should not be used up immediately because there is very little for bees to collect in European latitudes after about the beginning of December. No flowers bloom and it is so cold that flying is out of the question. The combs must now contain all the colony needs to survive the winter and to grow again in spring, because the colony's new year begins long before the spring flowers provide nourishment. The queen begins to lay again after the winter solstice, or at the latest with the first warm days at the end of February or beginning of March. Larvae hatch out of the eggs after three days and need carbohydrates, protein and fat for their development.

Here is where winter bees play their role. After they hatch, their first task is to develop well-fed bodies. This is not difficult because, as with humans, if they use little energy and eat 'normally', the result is a rounded belly. The winter bees have virtually no brood to care for, nor do they undertake strenuous foraging flights, because there is nothing left to collect. What they consume is stored as protein and fat reserves in their abdomens. Winter bees are chubby!

Cuddling up together

Cuddling for the duration of the winter break is not a bad idea. And round bodies are of course more pleasant to snuggle than skinny ones. So this is what bees do in winter – cuddle up together. Unlike other social insects, such as bumblebees and wasps, bees winter together as a colony. Bumblebee and wasp queens hatch in autumn and, after mating, leave their dying colonies. These young queens have a biological antifreeze in their tissues and hibernate hidden in the ground or under tree bark and establish new colonies in the following year.

Hibernation is unthinkable for honey bees. Instead, the females – and there are only females in the colony at this time – gather together in a winter cluster. When the temperature outside the hive falls, bees leave most of the passages between the combs and withdraw to occupy just a few of them, where they cluster together. With an outside temperature of below 6°C, the winter cluster takes on the form of a sphere. Depending on the size of the colony this may occupy five to eight comb passages, with most of the bees in the centre passage and fewer in the outer passages. The bees in the cluster keep moving and changing places. Those at the edge of the cluster burrow into the middle, forcing others out in a continuous turnover until it gets warmer and the cluster disintegrates.

Why the continuous turnover? Clustered together, winter bees keep the queen warm by vibrating their flight muscles with the wings uncoupled – like a car running in neutral.[1] Warmth is generated. Bees do not heat the entire hive, only each other, and move around continuously. The centre of the cluster is warmer than at the edges. Staying out on the edge would result in cooling down to the point of becoming immobile and falling off the cluster and dying. When all of them take a turn at the edge and then return again to the centre, they all survive. This behaviour can be watched live or on recorded videos taken by either normal or heat-sensitive cameras.

Bees are very sensitive to temperature, and winter clustering is highly effective. Single individuals become motionless at a temperature of 10°C and die at 4°C. An entire bee colony placed in a cold room can survive temperatures of −40°C and below, given two conditions: the cluster remains intact and they have access to their food supply. Vibrating their flight muscles to generate warmth requires energy provided by the carbohydrates in the honey. Should this be used up, they would soon freeze to death.

Bee colonies must therefore manage their food reserves sparingly, and studies of temperature in a winter cluster allow us to deduce how they employ their fuel reserves optimally. A thermometer placed within the passages between the combs allows long-term measurements to be remotely recorded and shows that the temperatures fall significantly both at the edges and also in the centre of the cluster to only slightly above 10°C. But every few days the cluster is well and truly heated up, climbing to over 30°C and then falling back to a lower level (Figure 2).[2] The strategy of intermittent heating clearly saves energy in comparison to keeping the cluster constantly at a high temperature all winter. But why not heat enough to keep it at a little above 10°C, warm enough so that they can all still keep moving? This would be even more energy-efficient. We are not exactly sure why, but one reason could lie with the consistency of the honey. At 10°C, honey is relatively firm and may be difficult to take out of the cells. A well-heated cluster would warm the combs and soften the honey stored in them. Days in the cluster when the heating is on could be party days, when there is something to eat. Bees feed one another, and perhaps on such days the bees on the honeycombs provide their sisters with their rations and allow them to survive until the next party day.

Figure 2

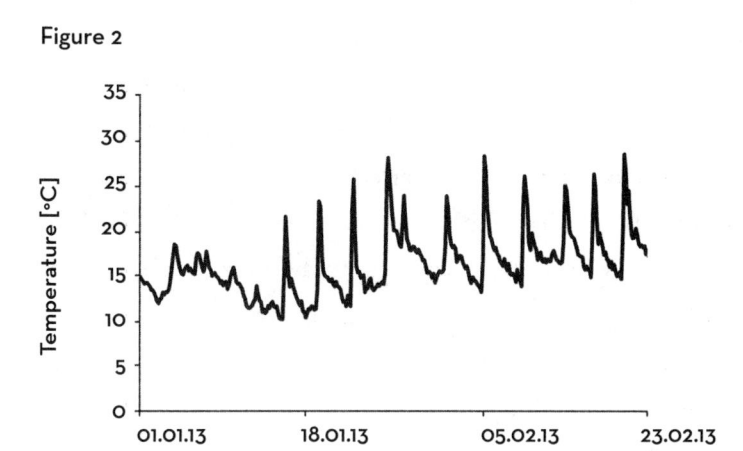

```
Temperature [°C]
35
30
25
20
15
10
5
0
   01.01.13   18.01.13   05.02.13   23.02.13
```

Temperature within a winter cluster in January and February 2013 (data from HOBOS project). Strong day-long heating peaks are clearly visible, between which the temperature sinks significantly. The gradual increase in average temperature over the displayed time is a sign that the queen began to lay about the middle of January and the colony had established a small brood that was being held at a constant temperature of 36°C (from Tautz 2015). Similar data for every winter since 2011 can be found via the HOBOS website.

The end of winter

The spirit of a colony revives after the winter solstice if it has not lost contact with its food supply and kept itself warm. The days become longer. At the end of February or the beginning of March, the temperature approaches 15°C and the call is: 'Stop cuddling and get outside.' The bees have an urgent need. In a hard winter, which even before Christmas was already too cold for outside flights, the bees have to sit on the combs inside the hive. The comb fulfils many functions, with one exception – sanitation. Wherever food is consumed and digested there is waste, and bees normally get rid of this during their flights out of the hive. When they are prevented from leaving the hive by the cold, there is only one alternative – hold on! Fortunately, nature has equipped bees with an expandable gut that can hold a fair amount. But eventually the pressure is too great and the bees start out from the hive on a 'purgative' flight. In times when there were no washing machines and driers, this may have led to tension in beekeeper households. One can imagine the carefully boiled, rinsed and wrung white linen hung out to dry on the very day the bees begin their purgative flights. The result, if one was not forewarned and could quickly bring the washing in, was a disaster. Hundreds of small, brown and unpleasantly smelling spots. Domestic responsibilities in those days were clearly divided between the sexes, so Mr Beekeeper could expect some unkind comments from his spouse on such occasions.

In European latitudes, the warm purgative flight days can be followed by cold spells, causing the bees to huddle together again in a winter cluster. Now they must keep not only themselves and the queen warm, but the first brood of the year as well. With the first warm days, a process begins that apiarists call the spring 'breakthrough' or the dwindling. The queen begins to lay eggs again and these will hatch three days after she deposits them in the cells. These larvae need a special diet to grow and to initiate the program of metamorphosis that will change them into bees. And now the hour of our winter bees has come. Like all worker bees, they can produce the special nutritional substance for the larvae from glands associated with their mouthparts. Under normal summer circumstances, bees that care for the brood, or nurse bees, use the pollen and honey collected by the foragers and process this into food for the larvae. In spring, there is scarcely any pollen and honey left in the colony, only carbohydrates from sugar. The fat and protein for the larval diet comes from the winter bees, who, in the previous autumn, had accumulated this in storage tissues in their abdomens. Now nurse bees, winter bees set about releasing these reserves.

Winter bee reserves are only sufficient for the very beginning of the new year and for the new brood. The colony must nevertheless grow rapidly in preparation for the new year. Consequently, the queen increases the number of eggs she lays and the bodily reserves of the winter bees are quickly diminished. Foraging must begin.

Just as well that willows begin to bloom from the middle of March. In good weather, winter bees can now be seen returning to the hive with heavy yellow packages on their hind legs. They are collecting pollen, and sometimes in such quantities that some combs in the hive are completely filled with the rich yellow protein. This is urgently needed. Many combs in the colony are now entirely filled with brood. Thousands of eggs and larvae need tending. The first generation of summer bees will soon hatch out.

This signals the end for winter bees. Sometime in April the last of them will start off on a foraging flight from which they will not return. An overabundance of newly hatched bees now occupy the hive; the spring breakthrough has occurred, the winter bees have passed on, and the time of the summer bees has begun.

Housework and child care

A young bee crawling out of her brood cell in early April has, like all her sisters that are born at the same time and in the following four months, a busy, short and strictly disciplined life in front of her. The beginning is also relatively tiresome. Scarcely out of her cell, her first chore is 'tidy your room'! And not only just your own room from which you have just emerged, but a thousand others that are part of the brood combs.

The first task of a young bee is to clean the comb cells in the brood nest. The cell left behind by an emerging bee is not exactly clean. Shortly before pupating, she would have emptied her gut and then spun a protective cocoon around herself, in which she metamorphosed and emerged from the cell as a worker bee. The cocoon and the adhered faecal remains are left in the cell. Cleaner bees take care of the situation. They clean out the cells with an oily substance secreted from glands around their mouthparts, after which the cells look like new. Beekeepers uncertain whether or not the colony has a queen are reassured if they find brood combs where the cells are clean and shiny. The cleaner bees have been at work and all is spick and span, germ-free and ready for the queen to lay new eggs in the cells. These must be tended and the larvae that hatch after three days fed and kept warm. This is the second responsibility taken on by bees in their lives. About four days after emerging from their pupal cells, they begin to feed older larvae with a pulp of pollen and honey. Six days after emerging, glands in their heads that produce a special nutritional secretion are fully developed and they can now supply young larvae with this protein-rich 'nurse's milk', depositing it in their cells. Larvae literally bathe in the milk and thrive. A newly laid egg weighs 0.3 milligrams. After nine days and just before pupating, the grub that develops from this egg weighs 150 milligrams – about 500 times more! Were human mother's milk to have the same effect, a nine-day-old baby would weigh about 1.5 tons.

Some nurse bees employed in the brood nest are required to take care of the most important individual in the hive. They are admitted to court and join the throng of nurse bees that constantly surround the queen, and their job is to groom her and lead her over the brood comb to lay eggs in the cleaned cells. Primarily, they must feed her. She does not get the same food as all the others: no beebread (the pollen and honey pulp) and no plain nurse's milk. Instead, she is fed with royal jelly. This queenly pudding, about which more later, is comprised of nurse's milk enriched with a high proportion of an additional secretion from mandibular glands, also located in the head and close to the mouthparts of worker bees. The queen receives this high-grade nourishment – which, incidentally, we humans do not find pleasant-tasting – for her entire life, enabling her to live up to five years and lay up to 5000 eggs a day.

The many young bees emerging from these eggs are certainly needed! A bee colony in European latitudes reaches its peak brood production between the middle of April and end of May regardless of the locality or weather conditions. This level is maintained until about the middle of June, when everything blooms. Honey factories must be prepared for these coming weeks if they are to enjoy a successful year.

The explosive increase in the number of bees in a colony that occurs in these weeks is impressive. Imagine the following: It is the middle of April. The spring breakthrough is nearly completed and the colony includes perhaps 6000 bees. The weather pattern is stable and temperatures lie around 15°C.

TEAMWORK IN THE HONEY FACTORY

Cherry trees begin to bloom and constitute the first large honey flow of the year for the honey factory. Large quantities of nectar and pollen are brought in to nourish the brood. Nurse bees, in the face of this plenty, encourage the queen to lay and she does the best she can, laying about 2000 eggs a day from around 15 to 25 April. Twenty-one days later, each of these eggs develops into a bee. In other words, 2000 young bees hatch each day from 7 March onwards. By 17 March, the honey factory will have not 6000 but 26,000 workers, and this number will nearly double itself. Bee colonies in the middle of June can incorporate up to 50,000 individuals.

Climate control in the nursery

Bees tend their many offspring with the same attention that many mammals lavish on their offspring. Bee larvae are cleaned, fed, and the pupae kept warm. Brood areas are maintained between 34°C and 36°C to enable larvae to hatch from the eggs, pupate and emerge as bees. How do bees control this process? A primary requirement is the ability to actively produce warmth by vibrating their powerful flight muscles, and as we have already heard, this is also how the winter cluster is heated. The same technique is applied to areas in the brood nest, but in a most interesting and unusual way.

Beekeepers that pull a brood comb out of the hive will sometimes see the tip of a bee's abdomen protruding from an empty cell in an area where most cells are closed. A 'heater bee' is at work here. With her head down in the empty cell, she warms the neighbouring cells and those on the side of the comb directly opposite. An empty cell serves as a heating cell over many minutes at a time if it is located in an area with many capped brood cells, whereas in smaller areas of capped brood cells the empty cells are not used for as long. Individual heater bees are usually active for relatively short periods. On occasion one may spend up to half an hour in a cell, although usually they change cells after less than ten minutes. If the comb is not warm enough when the heater bee leaves the cell, another takes her place. The first then looks for another empty cell or waits to be provided with honey (see Figure 3).

Heater bees burn a lot of fuel and heating brood cells is one of the most strenuous activities they will ever undertake. Oxygen consumption can be used as a measure of their effort. Bees that are actively heating use 1.14 µl/g of oxygen per minute. This is only slightly less than the 1.16 µl/g per minute they need to fly. Bees that heat without pausing for about thirty minutes use up their entire energy reserve of previously consumed honey. The exhausted bee finds it difficult to even reach the honey supplies, which are always some distance away from the brood nest.

Figure 3

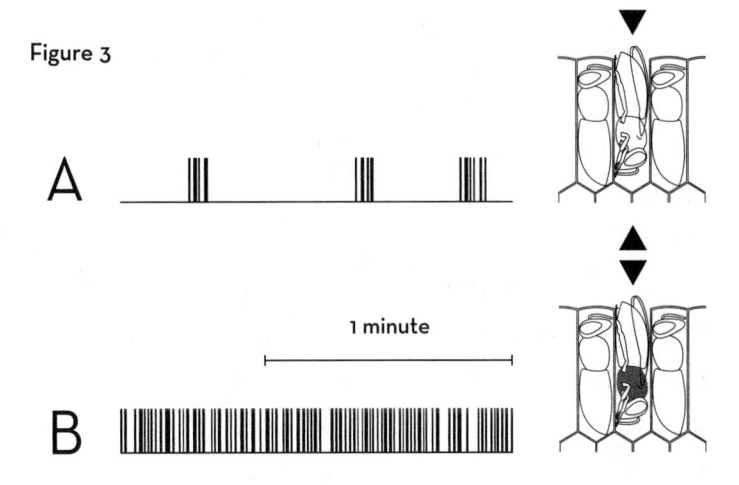

Two reasons why bees remain still over longer periods in empty cells. They are either asleep (A), when they breathe intermittently, shown by the pumping action of their abdomens filling the trachea of their respiratory system with air. Or they are heating (B), in which case their abdomens pump continuously. Each vertical line on the traces represents a single abdominal breathing movement. Bees must continuously replace oxygen in their bodies when heating. The behaviour described above was directly observed in a brood comb that had been very carefully removed from a hive (Bujok 2005).

Bees have solved this difficulty in a fascinating way. *Trophallaxis* is the term given to the bee's habit of mouth-to-mouth exchange of food. Analysis of several thousand food exchanges reveals that 85 per cent of these occur in the closed cell brood areas.[3] Two interesting details come from observing the donor and receiver partners of such exchanges. Heat-sensitive cameras determine that bees receiving honey always have a higher temperature than donors, leading to the conclusion that the receiver is a heater bee taking a break, and the heat-sensitive camera confirms this: receiver bees stay in the closed cell brood area and slip into the nearest empty cell to resume heating. Donor bees head towards the honey stores, where they pick up another load of honey and make their way back to the brood areas. Each of these commuter bees is a sort of 'filling station' bee, and she works as a team with heater bees to keep the brood nest at an optimal temperature.

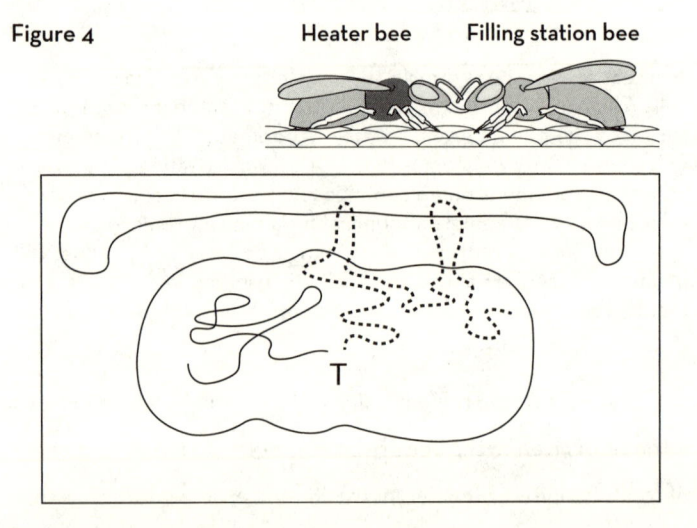

Figure 4   **Heater bee**   **Filling station bee**

Paths taken by heater bees and filling station bees. Heater bees (continuous lines) do not leave the brood nest, whereas filling station bees commute between the brood nest and the honey stores located above. Two bees met at point T, where the filling station bee (on the right) supplies honey to the heater bee (on the left), enabling it to continue with its activity (after Basile et al. 2008).

But what is the optimum temperature and how do bees establish where, when and how much has to be heated? While we do not know what the signals are that come from the cells surrounding a heater bee and activate and maintain her heating activity, there is good evidence that cues come from pupae. For example, a pupa can be removed through the base of her cell from the other side of the comb and replaced with a lump of wax without damaging the cap of her cell. This cell is no longer heated! Heater bees will also not attempt to heat an area of brood comb that has been carefully cut out, the pupae in it killed by cooling and then replaced in its original position.

However, heating behaviour is not determined only by signals from the brood. Heater bees can gauge the temperature of the closed brood cells with special receptor cells on the tips of their antennae. There are very few of these in comparison to the 20,000 or so receptors on the antennae that are used to detect odours. Most of the merely twenty or so temperature-sensitive receptor hairs are located on the endmost segment of the antennae. Here they are well placed to come into direct contact with their surroundings. Microsurgical removal of this end segment does not result in an immediate change in behaviour. Bees that have been operated on appear to be completely normal, although they have lost the ability to measure temperature, which is so critically important in brood warming.

This phenomenon was explored further in a special observation hive equipped with closed brood cells and populated with 400 bees, all without the end segment of their antennae. These bees did not heat the brood, which was to be expected because bees without temperature receptors are unable to tell if the brood temperature is too low. With no heating, the brood nest lies at about 24°C, the typical temperature of most areas in a hive.

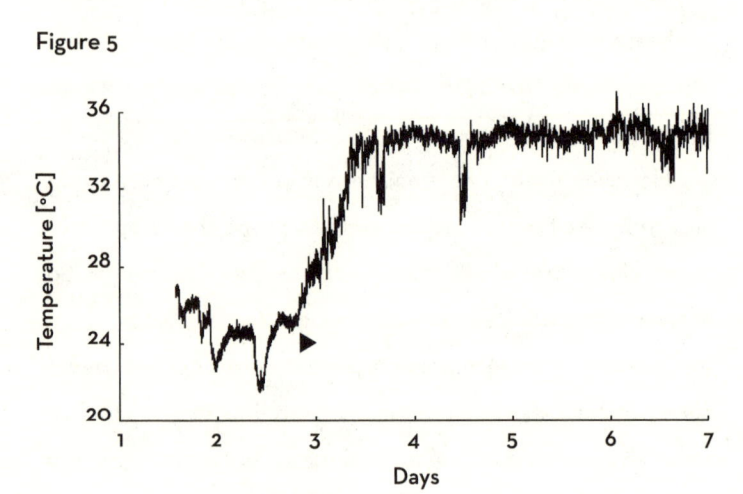

Figure 5

Air temperature in the brood area of an observation hive. Only treated bees were in the hive at the beginning and did not heat the area above normal hive temperature. After about two days, fifteen intact bees were added to the hive (arrow). Within a day, the temperature increased to the required level for brood areas of about 35°C (from Bujok 2005).

Fifteen worker bees with intact antennae, added to the 400 who had been ignoring the pupae, immediately began to heat the brood, again as expected. But to the surprise of the observers, the bees that had been operated on now joined in. They also activated their heating behaviour, and in twenty-four hours the brood nest temperature was raised to where it should have been.

Just how this collective heating behaviour is initiated is still a mystery. What signals do the bees exchange? One must assume that the same collaborative processes are operating here as in an intact colony, but these are not obvious to the observer because they are integrated in all the natural interactions between the inhabitants of a beehive.

The optimal warmth in the brood nest is not only the result of the bees' efforts and collaboration. The structure of the brood nest and the comb itself also play an important role here.

One of the most interesting aspects is the significance of empty and open cells scattered among the closed cells in the brood nest. Accurate measurements of heat transfer across the brood nest were made using small heating elements which were first installed in the thoraces of dead bees in place of flight muscles. The 'artificial' bees were then placed on and in the combs and heated to between 35°C and 45°C, equivalent to the temperatures that living bees can produce by vibrating their flight muscles.

Warm, living heater bees in the brood nest can be seen in three different attitudes: standing or moving freely about, pressing their thoraces firmly down against the cap of a cell, or with their heads down in an empty cell. Each of these three attitudes were simulated by the heater-equipped dead bees, which were left in place for thirty minutes – the longest period that heater bees remain in empty cells.

Figure 6

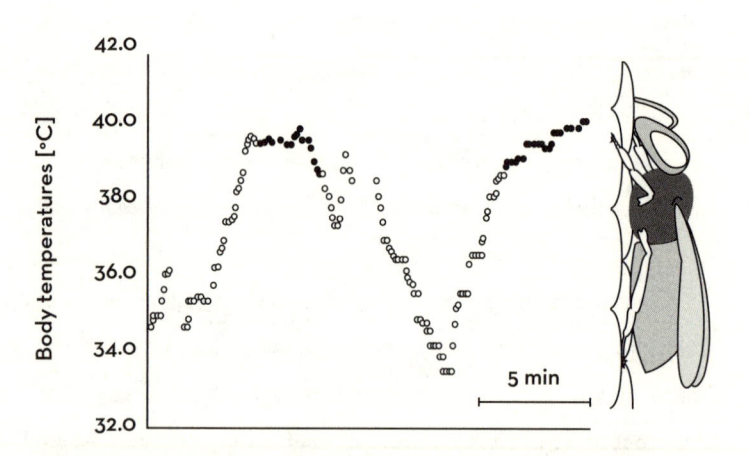

Regardless of which heating method bees use, they prepare for their effort by pre-heating their body temperatures. They warm themselves up (white symbols) before applying their thoraces to the cap of a comb cell (black symbols and drawing on the right), or before creeping head-first down into an empty cell to heat from there (see Figure 3) (after Bujok 2005).

Small thermometers inserted into the pupal cells from the other side of the comb so as not to damage the cell caps, measured the spread of warmth across the brood nest induced by the artificial bees.

The measurements show that the efficiency of brood heating is raised by a factor of three when a freely moving heater bee stops and presses its thorax down onto the lid of a cell. Heating from within an empty cell is the most effective method, and measurable changes in the inner temperature of a cell can be detected up to three cells away and on both sides of the comb. Thus, a single heater bee in an empty cell is able to warm more than sixty pupae! This perforated-comb central heating is the most effective way to heat the brood nest. Theoretical modelling indicates that a particular distribution of heating cells allows an optimal effect with minimal effort. Too few open cells lead to not enough heated pupae, which also store the warmth. Too many open cells, for example in a diseased colony or with a queen that is not laying enough, significantly increases the effort required to heat the brood. Quantified, the energy required to heat a comb containing 4 to 10 per cent open cells is minimal in comparison to a completely closed comb, and a structured comb saves the heater bees up to 40 per cent of heating time.

Some cells in the brood comb always remain empty, because the queen does not lay her eggs in a perfectly systematic way in all empty cells or on the same side and region of the comb. A pattern of holes therefore develops in every brood comb and is often favourable for saving time and energy for heating the brood area. Do the bees structure their brood nest according to a plan? We do not know.

What we do know, though, is that neither the effort of the bees nor the structure of the brood nest is alone responsible for the warmth enjoyed by developing bees. The comb also functions like a greenhouse. Heat diffuses through the comb, partly by conduction through the wax, through air within cells and through the bodies of bees and their brood. Heat is also transferred by radiation. Hot bodies radiate warmth to their surroundings. We perceive such thermal radiation when we hold our hands close to room heaters to see if they are working. Thermal radiation directed at material that is partly permeable to it will not all be reflected. The amount that is transmitted depends on the material itself and on the wavelength of the radiation, which in turn is related to the temperature of the source.

Together these lead to the so-called greenhouse effect. Thermal radiation directed through permeable walls of an enclosure and onto an object within the enclosure will warm it. This now radiates in turn, but its radiation may be at a different frequency from the original source, and cannot be transmitted out of the enclosure – this leads to thermal radiation being transmitted in only one direction. The initial radiation and warmth are trapped. Radiation from the sun as a source of warmth passes through our atmosphere to the earth (and can be trapped there!). A similar process occurs in the brood nest of bees. Heater bees are sources of thermal radiation, which passes through the thin cell walls and falls on pupae in neighbouring cells. They absorb the warmth and in turn become sources of warmth, albeit at a temperature lower than the heater bee.

The transmission of thermal radiation through the wax walls of a comb has a complex profile.[4] The thin beeswax walls present little impediment to some wavelengths in the relevant temperature range, allowing the radiation from heater bees to pass through. Radiation with wavelengths close to these are blocked by the walls. The lower temperature of pupae, together with the physical properties of the wax, is the basis of the greenhouse effect in brood nests.

Bee-bionics in wax and combs

The enormous brood production of a colony needs more than clever and efficient use of energy. And there is another somewhat simpler problem – where to accommodate all the brood? The comb area has to be expanded to provide more brood cells, and the life cycle of summer bees is adapted to this need. Summer bees spend their first ten to twelve days in the brood nest cleaning and tending the brood of their colony. On day twelve, their life changes and they go into comb production.

About ten days after emerging, all hive bees have fully developed wax glands on their abdomens from which they can extrude small plates of wax. Their colleagues take these, chew them to render the wax more pliable, and build the wax into combs.

Beekeepers give their colonies more space when they begin to grow with the onset of spring flowering. Honey factories receive another floor with the addition of a new case, or super, and frames on top of the existing hive. A few frames without the central foundations are also included. Some of the young bees in the hive now start work as builder bees and the new frames are occupied by building gangs. Working close to one another on both sides of the comb, builders draw up cell walls and provide space for more brood and honey. Strong colonies can produce up to ten completed combs in less than a week if fruit trees are in full bloom, the weather is kind and the honey flow is good. How do the bees manage this?

The desire to understand and explain how honey bees construct their combs is driven not least by their unbelievably precise appearance – an area of perfect hexagons with paper-thin walls built from a special material. Similar structures are found among species related to bees: the brood cells of colonial wasps, for example. These are also six-sided, but made from chewed cellulose and not nearly as precise as those of bees. Bee combs are so accurately formed that astronomer and mathematician Johannes Kepler (1571–1630) attributed bees with an understanding of mathematics to explain their achievement. René-Antoine Ferchault de Réaumur (1683–1757), a French natural scientist, suggested using the dimensions of bee comb cells as a standard unit of measurement for length to resolve the prevailing chaos of feet, cubits, spokes and spans, which all meant something different in different villages. However, such recognition of bees' building skills was denied them when, on 26 March 1791, the National Constituent Assembly in Paris decided, on the advice of the Academy of Sciences, to adopt one ten-millionth of the distance from the equator to the north pole as the standard unit of measurement for length. And so we measure length in metres, rather than bee cubits or something similar. Which is just as well, because in fact the dimensions of the regular, fine and stable structure of bee combs do vary slightly between colonies.

We do not know exactly how the regularity of the combs arises. Observing a single bee in the turmoil of building crews reveals how they work on the cell walls with their mouthparts. Chewing, plastering and pushing, builder bees raise the cell walls up from the central partitions of the frames. The walls vary in thickness by only a few thousandths of a millimetre along their lengths, and the surfaces are completely even. Computer-controlled machines can do no better with such fragile material.

Several theories have been advanced to explain this phenomenon. One of these is the following: each bee works on one side of a wall unaware of what is being done on the opposite side. Both builder bees press against the wall with their feelers. They have an instinctive conception of how pliable the wall should be when it has the correct thickness. By continuously scraping the wax from the wall and testing it, according to the theory, they eventually produce walls an optimal thickness and great regularity.[5] More recent research using new techniques points in a different direction and has led to the development of a theory based on precise measurement and not on appearance alone.[6]

Accurate measurements of the temperature of small areas within the colony, taken with heat-sensitive cameras, has provided us with a much clearer picture of the distribution of different temperatures throughout the comb construction area. Such techniques have led to completely new insights into comb building.

Images from a heat-sensitive camera focused on a building area in a natural hive clarify several aspects of the building process. Newly built cells lie close together at the edge of the comb.

Initially, these are cylinders that bees shape around themselves using their bodies as templates. Heater bees then slip into the cells and bring the walls and floors up to a temperature of over 40°C. Wax at this temperature almost flows, so that the vertically drawn up cylinders fuse and a physical process comes into play, resulting in an exactly hexagonal structure.

Hexagons arise in both the material and natural worlds wherever evenly distributed forces work against one another. Mechanical tension within comb cell walls pulls the flexible wax into line, and the sides of the closely packed cylinders coalesce into cells, each with six straight walls. Their surfaces are completely smooth, have a uniform thickness of 0.07 millimetres and form an angle with one another of precisely 120 degrees. This fascinating and precise structure results from the interaction of the physical properties of beeswax and the bee's ability to generate heat.[7]

Had Johannes Kepler seen thermal images of comb building he would have hardly credited bees with mathematical talent. He would have had to do the same for soap bubbles, for they exhibit the same phenomenon. When two soap bubbles touch one another, a wall forms between them that is perfectly smooth and has an even thickness. Incidentally, the strategy of forming exact hexagons by heating and fusing round apertures in suitable materials can be found in a recent technical patent.[8] Bee-bionics are moving into the world of humans.

Honey makers

Bees that build combs have to 'sweat it out' twice. First, to produce building material. and then to heat it up to a temperature at which it fuses to take on its correct form. Do summer bees that do not join builder crews have it any better? On the eleventh day of their lives, they too enter production in a honey factory. Their business is pollen and nectar. They do not have to 'sweat it out'; instead, their job is to stand close to the hive entrance to receive what foragers bring in, because foragers do not store their precious cargo themselves. Honey makers take the pollen bundles from foragers in their mouthparts and either carry these directly to brood combs, where nurse bees feed it to the brood, or they collect and store it in pollen combs by pressing it firmly down into the cells with their heads.

Foragers with nectar transfer this to honey makers, who store it first in a comb located in the lower part of the hive, the 'ground floor'. Other honey makers now take the nectar from there and carry it up to the next floor, where yet another crew of honey makers accept it. The nectar eventually finds its way from the ground floor up to the 'penthouse' – the honey floor.

Do honey makers have it easier than builder bees? Maybe. A single bee can carry only 50 to 60 milligrams. To store 100 grams of nectar requires the contributions of 2000 foragers, but on a fine day in spring a colony can collect up to 3 kilograms of nectar! Such days are stressful for honey makers. Foragers crowd at the entrance, waiting to deliver their 60,000 individual loads, all of which are carried about the hive by honey makers and exchanged several times. Builders may have to sweat it out, but honey makers have to run. Neither have it easy.

Another change occurs in the bodies of summer bees while they are busy with building or nectar processing, preparing them for their next assignment. The venom glands of stings fill up and the production of scent, used to signal their sisters, increases. The summer bees are preparing to leave the protection of the hive and the circle of caring colleagues. They go to work outside and are no longer hive bees, but foragers or field bees.

Guard bees

Some field bees do not immediately go out into the wide world. Between their eighteenth and twenty-first days they take up positions at the hive entrance as guards and check incoming bees. Should they be admitted? Are they friend or foe? Scent provides the answer. Hive sisters have the same body odour as guard bees and are let in. Those who do not smell right have a problem. Bees from a different hive, for example, will be threatened and pushed away should they try to pass, and in most cases they realise their error and fly off. If they persist in their attempts, guards employ their stings and attempt to kill the attackers, in which case the bees grapple with one another on the landing board, each trying to be the first to sting. A guard bee has the advantage, in that she can call for help if she is not able to subdue the foreigner. She releases an alarm pheromone (scent) to alert her sisters, and it can then go badly for the rival, who is best advised to leave quickly.

Such scenes are rare, because attacks on other hives seldom occur. Also, bees know the way to their own door very well and do not often make mistakes. Beekeepers refer to field bees being 'flown in', somewhat like aeroplanes following a guiding beam, directly to the landing boards of their hives without having to search for it. Bees are not able to recognise the correct hive boxes by their appearance, though, only the place and position of the hive entrance where they started out and to which they return. Beekeepers make use of this when strengthening a failing colony with field bees from a stronger colony nearby. The hive of the stronger colony is moved to the side on a sunny day when most of the field bees are out and about, and the weaker hive is put in its place. The returning bees do not waste much time. There, where their landing board previously stood is another and they march straight in. To begin with, the guard bees are naturally somewhat excited. Masses of strange, rough and rude field bees arrive at the entrance, but their sheer number prevents any serious resistance from even starting. Guard bees can also be bribed, and entering field bees bring nectar with them in their honey stomachs, which they willingly give to the honey makers standing by.

How 'flying in' comes about is not quite clear. At times, about noon on a warm day, an interesting phenomenon can be observed in front of some hives. Large numbers of bees fly out of the hive and buzz back and forth around the entrance. When several colonies are located close to one another, the buzzing can be heard from many metres away. For a long time, it was believed that this behaviour represented the 'play flights' of young hive bees preparing to enter service as field bees. It was thought that these trial flights enabled young bees to assimilate an image of the hive entrance and surroundings, allowing them to find it after a foraging excursion. However, catching bees from such a buzzing throng revealed most of them to be experienced field bees. The 'play flights' appear to have little to do with the development of the flying-in ability. Meanwhile, there are other explanations for this behaviour that we will return to later on.

Perhaps young field bees learn the way home simply by flying with experienced foragers. Radar techniques, employed to record the first flights of individual bees from hives, show that they fly out from the hive and come back along the same path. Taken together, tracked flights form a star-shaped pattern in the air around the hive. Are they laying a scent around the hive? We have no idea. Also, only the flight paths of single individuals have been tracked so far and it is not known if the first flights of all bees are the same. New young foragers may be following scent trails left by experienced foragers. Scent does play a very important role in orientation for bees and there is a special unit of field bees whose exclusive assignment is to lay scent trails.

Scouts

Honey factory scout bees are like prospectors that work for companies that mine ore or drill for oil. Scouts search for new sources of raw material that the colony needs for its development and survival. If there is a shortage of water, scouts set out to find a supply. Should the colony need pollen, then they seek new sources of pollen, and if new accommodation for the hive is required, scouts busy themselves as estate agents, as we will see later in more detail. But the primary task of scouts is to find profitable sources of nectar, the most important raw material for honey factories.

Scout bees are individualistic and are the first to leave the hive each morning. They fly alone, often far from the hive, and can be found within a radius of the hive of up to 5 kilometres. Scout bees land on flowers and check the nectar content. They gain an overview of how many flowers in the area contain the same nectar and how far from the hive they lie. Is the nectar supply large enough to be worth visiting? Would foragers bring more energy into the hive than it would cost them to fly there and back? When they have an impression, scout bees mark the flowers with scent, gather samples of nectar and pollen and fly back to the hive. Here they inform the foragers in the hive by performing a unique 'waggle dance' which conveys the location of the source, the kind of flowers from the nectar probe, and the kind of scent and pollen adhering to their bodies. Armed with this information, recruited foragers set out.

Forager bees

Most field bees are foragers that set out from the hive with information about the location and nature of the nectar source. Arriving at the site, they are led to the goal by the scent trails of both scout bees and any colleagues already there. Once their honey stomachs are filled with nectar, the foragers fly back to the hive, where the honey makers are waiting. Some foragers collect pollen and others gather propolis. More about propolis later. Summer bees carry out their foraging duties for three to four weeks, and during this time the young cleaner and nurse bees have become experienced and seasoned field bees. As young bees, they possessed a furry covering on their bodies, but now they are bald from the continual bustling against the bodies of sisters in the narrow confines of the hive. Peaceful and somewhat simple-minded as young bees, they are now worldly, equipped with filled venom glands in their stings and not always friendly. Beekeepers prefer to work with their hives in spring and summer around noon on a warm day, because colonies are then often quiet and peaceful. The usual explanation is that most field bees are out and away at this time: the aggressive Amazons are not at home. Is this really true?

Honey bees and sleep: Too tired to attack

In good summer weather, forager bees start their working day at 6.00 am and end it at 8.00 pm. Records of flight activity at the hive entrance during this time show something unexpected. Around noon in summer there is a period when departures and landings are significantly low. Apparently, many bees take a midday rest after a busy morning. Are they napping?

One thing is clear: bees must sleep. Honey bees are asleep when their antennae sink down, muscle tone decreases and they breathe more slowly (see Figure 3). Martin Lindauer, one of the founding fathers of German bee research, drew attention to 'sleepy' bees many years ago[9] and a publication by Walter Kaiser opened the way to the systematic study of sleep in honey bees.[10] Subsequently, scientists found that when honey bees are deprived of sleep, their memory, communication and

Figure 7

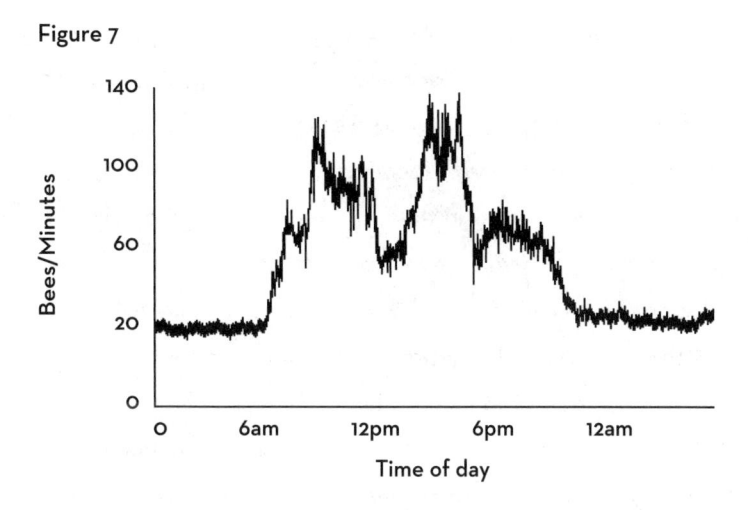

140
100
60
20
0

Bees/Minutes

0   6am   12pm   6pm   12am

Time of day

Flight activity (number of entries and exits of foragers per minute) at the hive entry on a typical summer day, here 3 June 2013 (from the database of the HOBOS project). It is relatively quiet around 1 pm.

concentration abilities suffer.[11] Bees with a sleep deficit have trouble finding their way to food sources and back to the hive, and their dances are imprecise.

How much sleep does a honey bee need? There is no universal answer to this question because the needs of bees depend on the tasks they undertake in the hive and on their maturity. The sleeping patterns of honey bees are specific to each activity group. Workers, as we have described, have several vocations during their lives, first within the hive and then outside of it.

A worker is first a cleaner bee, then a nurse bee, later a honey maker and finally, in the last phase of her life, a forager in search of food. Their sleeping behaviour in each of these phases is as different as the worlds they experience.

Young worker bees in the hive take many small breaks of a maximum of thirty minutes over twenty-four hours. There is no rhythm of daylight and darkness in the perpetually dark hive, so they can sleep at any time. In twenty-four hours they will sleep for a total of four to six hours. Most find empty cells in which they can bunk down. Endoscope cameras recorded novel sleeping postures, in which bees wedge themselves between two combs, supported by their heads and abdomens, and let their antennae and legs dangle down. Of all bees working within the hive, the honey makers that receive foragers' loads sleep longest.

Honey bees that shift from working inside the hive to outside it change their sleeping patterns and quarters. Foragers sleep at the edge of combs, where they are not continually jostled by their nest mates and find some peace and quiet.[12] Older foragers sleep well. Their bodies cool down and their sleep is deeper than other bees. In contrast to hive bees, they develop a day–night rhythm because they are exposed to a daily light–dark cycle, unlike hive bees, whose workplace is always completely dark.

Lazy bees

Bees are not only sleepyheads, they are also relatively lazy. Standing near a hive and watching the flights of diligent foragers one is inclined to believe that they are all 'busy bees', but this impression does not stand up to close scrutiny. Theodor Weippl, director of the apiarist school in Vienna in 1928, was interested to know just how busy the honey bees really were. The results of his observations and calculations were published in the *Archiv für Bienenkunde*.[13] Using the amount of honey and pollen consumed and the quantity of wax produced as indicators, he calculated that over a time span of twenty-one days, foragers averaged 2.8 flights per day.

This is not a particularly impressive manifestation of hard work! Did Weippl miscalculate? Could it be true that the busy young ladies from the hive are in reality just lazy? Weippl's results were viewed sceptically at first, because in experiments single bees can be induced to make many more foraging flights. Is the foraging behaviour under such experimental conditions identical with that of worker bees in a natural and freely foraging colony? Critical information was provided by the following experiment: A colony of 4000 bees was established in an observation hive and marked foragers observed from 5 am to the end of their activity between 7 and 8 pm. The percentage of foragers that undertook foraging flights and how often each forager left the hive was recorded. The observations took place over six three-day periods, evenly distributed through the months of May, June and July.[14]

The results confirmed that, depending on the weather, the food situation in the hive and the availability of nectar sources, between 0 per cent and 70 per cent of the colony's bees undertook foraging flights. Focusing on fifty randomly chosen individuals produced a surprise. The average number of flights per day was only 3.4. The most industrious bees completed up to ten flights per day; the least industrious only one. The majority took off four times each day. Does this look like hard work?

The situation appears in a somewhat different light when we consider these results in terms of the somewhat moderate foraging activity of individual bees. Let us grant that a strong colony in summer can harvest a record of 5 kilograms of nectar in one day. We assume a colony of 50,000 bees, which is not unusual in summer. Third, let's say that half the bees on the day in question undertake foraging flights. Each bee is able to return to the hive with 50 milligrams of nectar for each flight, allowing us to calculate how many foraging flights must be made to bring in 5 kilograms. Each bee must fly twenty times for each gram, and 100,000 times for 5 kilograms. With 25,000 foragers, this means just four flights daily for each bee. The experimental observations provide very similar values. With just this amount of effort, a colony actually achieves a record for nectar harvesting!

Additional evidence has come from researchers who attached small RFID (radio frequency identification) chips, like rucksacks onto the backs of worker bees as they emerged from their pupal cells. With these devices, the number of times individual bees left the hive could be automatically recorded. Here too, between three and ten flights per day was the normal value for individual foragers.[15]

Bee colonies do not achieve their amazing performance in terms of nectar harvesting because individual foragers are so busy. Instead, bee colonies are an example of how effective teamwork can be, even if to some, teamwork means 'good, someone else will take care of it'.

The waggle dance: Rethinking old concepts

When questioning the diligence of bees, it becomes clear that some issues turn out to be very different from first impressions. New methods of observation and analysis lead to new data that do not fit with the initial image of the truth. But it's not only new data that leads to altered judgements. Older data can be completely reinterpreted if prevailing ideas are ignored. This is the case with the 'dance communication' of bees. We have already mentioned how important this form of communication is for honey bees. Scouts use dance communication to inform foragers in the hive about profitable food sources. But exactly how does this communication work?

Usually, the answer to this question is based on the Nobel Prize–winning studies of Karl von Frisch (1886–1982). Scouts and foragers that have found nectar provide information about the direction and distance of the source from the hive in a 'waggle dance'. Followers of the dancing bee can then find this source using information they have gathered from the nature of the dance movements. In the words of von Frisch: 'Foraging bees in contact with a good food source rapidly bring new recruits to both distant and hidden locations. They are not led to the site, but sent. The waggle dance informs them about the direction and distance.'[16] School children are often shown a simple graphic representation of this process (Figure 8). But does this simple model really depict the complexity of communication in a bee colony? Let us first examine how such a model was derived.

Karl von Frisch observed that foragers returning from a food source performed remarkable movements in which they walked slowly forward, while waggling their abdomens rapidly from side to side. After a short progression, they turned alternately either left or right, then circled back to the start without waggling before commencing another forward waggle run. The path

Figure 8

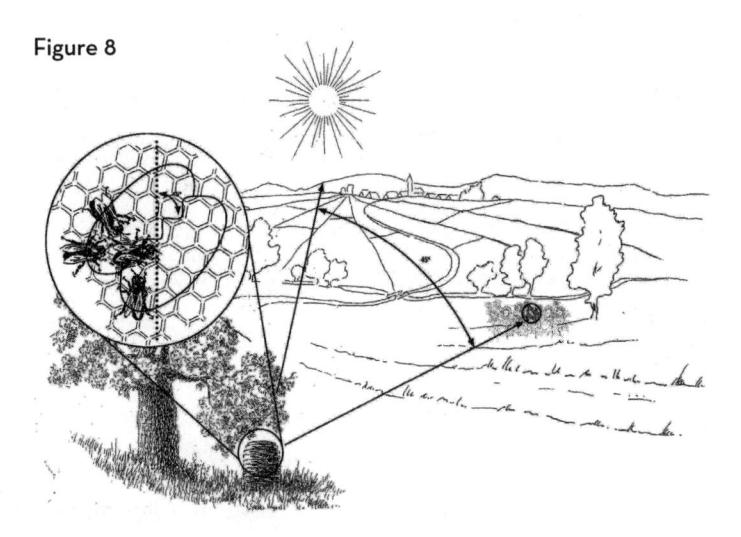

The well-known description of the dance language, after Karl von Frisch (first published in this form by von Frisch, 1965, p. 137). The figure traced in the dance expresses the direction and distance of the feeder from the hive (from Tautz 2015).

they follow resembles a figure eight on its side. Von Frisch called this the 'bee dance' and believed it could serve to communicate locational information. Two important observations led him to make this assumption: (1) Recruited bees soon showed up at the place where the dancer had collected nectar. (2) The forager's dance movements altered depending on the location of the food source in the field and, for stationary sources, with the movement of the sun, which evidently served as a reference point for the dance movements.

Von Frisch tested his idea that bees were using the dance to convey spatial location with his famous 'step' and 'fan' experiments. The aim was to determine whether the recruits indeed followed the instructions contained in the dance to reach a goal. Step studies were designed to show if information about distances was conveyed; fan studies to see if directional cues were followed.

What was the result? Let us consider the step experiment of 3 September 1962. A feeding station – a small bowl containing sugar water and made more attractive with scent – was set up 300 metres from a hive (F in Figure 9). Seven foragers marked with coloured spots of paint were trained to visit the feeder. These bees recruited eighty newcomers in 2.5 hours, all of which were caught at the site. Seven additional control stations were set up along a line between the hive and the feeding station. These were scented but contained no food. Newcomers also arrived at these stations during the 2.5-hour observation period. From these results, it became clear that: (1) More bees landed at the control station closest to the feeder at F than at the others. (2) Twice as many recruits landed at the feeder the foragers had visited than at all the control stations together. These observations raised questions concerning the precision of the locational information experienced foragers give to followers. To find an answer, we have to take another careful look at the waggle dance.

Figure 9

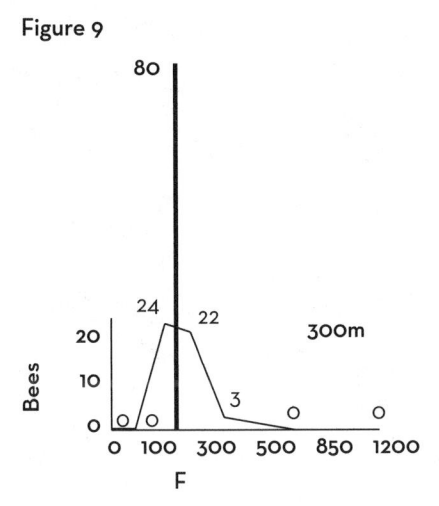

```
Bees
     80 |
        |
        |
        |
     24 |
  20    /\  22           300m
  10   /  \
        |   \  3
   0  o/o   \____o_____o
      0 100 300 500 850 1200
      F
```

A diagram from Karl von Frisch's book (von Frisch 1965 p. 94) shows how many recruits appeared at empty control feeders. The original figure has been expanded here (solid vertical line) to show the number of recruits that landed at F, a feeder to which experienced foragers had been previously trained and had visited.

How accurately can dance movements be measured?

To determine if a relationship existed between the dance pattern and the location of a goal, von Frisch and his co-workers measured the angular direction of the waggle dance in relation to vertical. An extract from the original protocol of dance measurements in which angles are recorded with an accuracy of 0.5 degrees is shown in Figure 10. However, from one look at the dance movements recorded with the presently available optical methods (Figure 11),[17] it is clear that the direction of dance paths cannot possibly be determined with an accuracy of 0.5 degrees. The dance pattern shown in Figure 8 is therefore highly idealised. How can such an exact waggle dance angle be drawn from the real paths traced by dancers. We have to conclude that the values recorded by von Frisch were estimations rather than exact measurements.

Figure 10

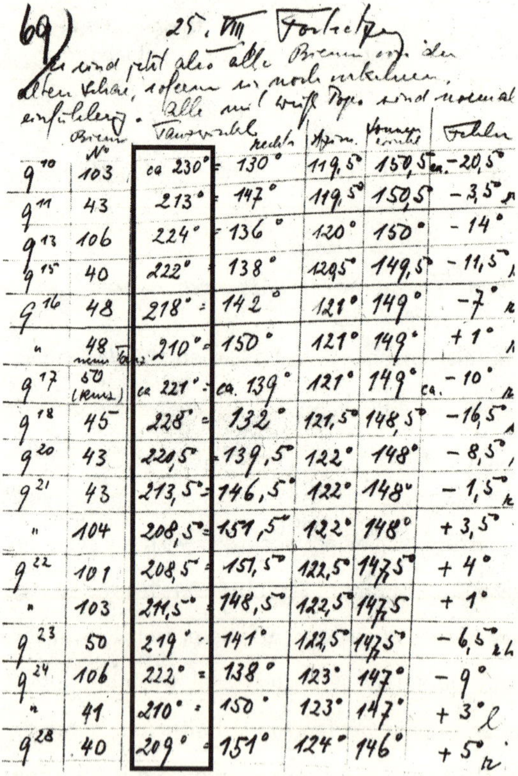

9^{10}	103	ca 230°	130°	119,5°	150,5°	ca. −20,5°
9^{11}	43	213°	147°	119,5°	150,5°	−3,5°
9^{13}	106	224°	136°	120°	150°	−14°
9^{15}	40	222°	138°	120,5°	149,5°	−11,5°
9^{16}	48	218°	142°	121°	149°	−7°
"	48	210°	150°	121°	149°	+1°
9^{17}	50	ca 221°	ca. 139°	121°	149°	ca. −10°
9^{18}	45	228°	132°	121,5°	148,5°	−16,5°
9^{20}	43	220,5°	139,5°	122°	148°	−8,5°
9^{21}	43	213,5°	146,5°	122°	148°	−1,5°
"	104	208,5°	151,5°	122°	148°	+3,5°
9^{22}	101	208,5°	151,5°	122,5°	147,5°	+4°
"	103	211,5°	148,5°	122,5°	147,5	+1°
9^{23}	50	219°	141°	122,5°	147,5°	−6,5°
9^{24}	106	222°	138°	123°	147°	−9°
"	41	210°	150°	123°	147°	+3°
9^{28}	40	209°	151°	124°	146°	+5°

A table from the original experimental protocol of Karl von Frisch (from Kreuzer 2010).

We find a similar situation when considering the accuracy of bee communication about the distance between the hive and the feeder. The further away the feeder is set up from the hive, the more similar the dances become, although the actual distances vary considerably. By no means is a particular distance coupled with a set number of abdominal movements. Instead, it's more common to see waggle runs at first increase significantly as the feeder is moved further away, but with increasing distance between hive and feeder their duration alters less and less. There is, therefore, no linear relationship between the length of the waggle run and the distance flown. The information provided by bees covering long stretches is increasingly inaccurate for distant goals, becoming more and more similar for very different distances flown. And that is not all; there is another complication.

The duration of the waggle dance, or number of waggle movements that a dancing bee expresses, does vary when the distance between hive and feeder changes. But as explained above, this becomes less accurate the further away from the hive the feeder is. Perhaps at least this inaccuracy is consistent, regardless of which areas foragers visit on their foraging tours? Sadly, this is not at all the case.

Bees employ an optical odometer, or distance measuring system.[18] They judge the distance between the hive and their goal using the visual perception of the landscape through which

Figure 11

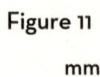

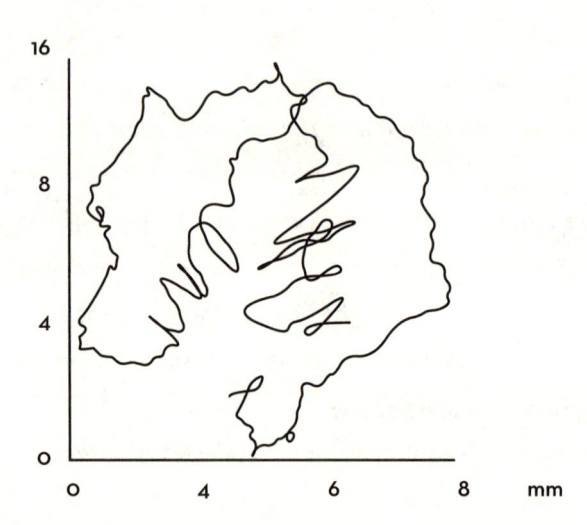

Precise recording of the dance pattern (the path traced by a dancer on the comb) of two sequential rounds with which a dancer advertised a goal 215 metres away from the hive (Landgraf et al. 2011).

they fly.[19] But such a method is strongly influenced by the structural nature of the landscape and is inaccurate. Twenty waggle movements can mean a distance of 80 metres if the bee flies through a highly visually structured landscape. Twenty waggle movements could also signal 180 metres if the flight passed over a low-contrast field with no landmarks on the horizon.[20] Suppose we have two dancers from a hive facing a country road. A wheat field lies on one side of the road, and beyond it an orchard with blossoming trees, 180 metres from the road. A farm with several buildings stands on the other side of the road and a similar blossoming orchard beyond that, only 80 metres from the road. One of our dancers flies to an orchard across the visually plain wheat field. The other flies past the visually diverse farm buildings to her orchard. Both report rich food sources and do so with the same waggle frequency, although the bee flying over farm buildings has only flown 80 metres, compared to 180 metres flown by her colleague. Both 'think' and report that they covered the same distance.

There is yet another difficulty. Two dancers that fly over the same stretch between the hive and the same feeder and dance to report their findings do not carry out even close to the same dance movements. Karl von Frisch also noticed this, and in a publication in 1957 remarked, 'Bees sent out by dancers fly to the goal with a greater precision than would be expected from the variability [of the dances].'[21] His explanation for this phenomenon: 'One can conclude from this that while following the dance, bees take an average of the individual values.' Is this not perhaps expecting a little too much from bees? Is it plausible to suppose that recruits follow many dances knowing that the instructions given by their sisters are not exact and then calculate an average that will lead them to the goal? And this still does not take into account the other inaccuracies described above.

Karl von Frisch developed a model that contributed a great deal towards explaining how bees communicate the location of food sources to one another. However, the model does not provide as conclusive an answer to the question as he thought, although it is still accepted by many to this day. Modern radar techniques have allowed us to establish that after following a dancer in the hive, bees fly off in a broad directional fan and not as though drawn along a thread to the feeder.[22] An experiment showing that the formation of the fan arises from the foragers' own dances is documented in Figure 12. Twenty foragers from a hive (H) were trained to a feeder (F) 248 metres away. The dance figures of foragers were recorded with modern video techniques and analysed following the methods of von Frisch. The experiment analysed 1300 recorded flights from the twenty foragers. Instructions contained in dances directed foragers to a large area, with 90 per cent of all dances indicating locations within a certain perimeter in Figure 12. Only 15 per cent of the dances indicated locations within a radius of 50 metres around the feeder. Bee dances do not indicate a precise goal such as a tree. Instead, they point to a general area where the tree stands. So how then, do bees that reach the general area of the nectar source find its precise location?

Figure 12

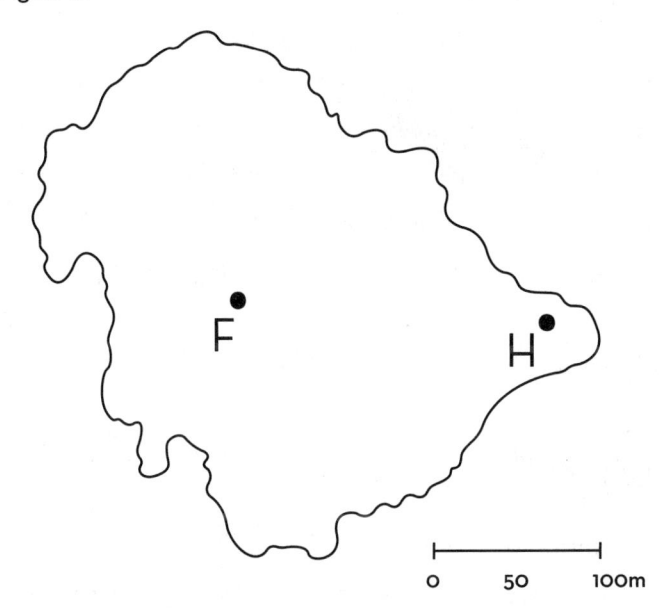

Large numbers of foragers were trained from the hive at H to the feeder at F. A goal was calculated according to the classical model for each and every dance-round performed by bees. The solid line encloses a perimeter in which 90 per cent of the more than 1000 calculated goals lay, which were indicated by dancers (after De Marco et al. 2008).

Follow your nose!

To answer this question, we can again turn to an observation Karl von Frisch made long before he developed his dance model. His opinion at that time was that the presence and activity of bees familiar with the goal was critically important for successful recruitment. It occurred to him that bees that had already flown to a site drew attention to themselves by performing so-called 'buzzing flights' around the feeder. During these flights, bees extrude pheromone-producing Nasonov glands, release the pheromone geraniol and establish an attractive lure in the air around the site. A table recording von Frisch's observations[23] shows that with 'rich' feeders, forty-six of sixty-four arriving bees had visibly extruded scent organs and performed buzzing flights. In contrast, none of the thirty-three bees arriving at a 'sparse' feeder performed buzzing flights. The coupling between dances and buzzing flights was recently studied by a group of scholars taking part in a youth research competition in Germany.

Karl von Frisch explored the effect of buzzing flights with extruded scent glands on recruiting success in a series of experiments. Two groups, each with seven foragers, were trained to fly to two separate feeders placed opposite one another and set up 10 metres away from the hive entrance. The scent organs of one group were glued closed, and those of the other group left intact. About ten times as many recruits arrived at the feeder marked by intact bees with buzzing flights, compared to the site visited by bees with closed scent glands. Von Frisch concluded, 'With this it is demonstrated that the normal strong stream of newcomers to a feeder is mainly determined through the extrusion of scent organs by foragers.'[24] In the early stages of his research, von Frisch emphasised the clear relationship between bee dances and the scenting of the goal. He remarked, 'Only after returning to the hive, dancing, and then flying again to the feeder do they extend their scent organs.' Furthermore, 'Foragers from the hive arriving at full feeders often buzz around it in irregular circles for a noticeably long time before they settle ... I have only now noticed that during the buzzing flight the scent organs are mostly extended and hence thoroughly impregnate the rich food source with a special scent ... In general, the longer a bee has danced on the comb, the more thoroughly she scents the area on her return to the feeder.'[25]

Decades later, von Frisch distanced himself from his discovery of the interaction between the several goal-finding cues. He purposely ignored the importance of events at the feeder in his step and fan experiments, focusing entirely on the success of dances in directing recruits. He commented, 'Naturally the stream of newcomers arriving at exactly the location indicated by the dance, namely a feeder, is of particular interest. The arrival at a feeder, though, must be left out of the analysis of the results. Completely different conditions prevail at a feeder compared to those at an observation plate, due to the scenting activity of marked foragers at a feeder. The additional attraction of scent varies from experiment to experiment and depends strongly on its intensity and wind direction.'[26]

Send them out, then draw them in

We arrive at a broader view of the recruitment of honey bees to feeders if we accept all that von Frisch established in his thorough studies and careful experiments but include, instead of disregarding, his observations of dances and buzzing flights. Karl von Frisch had already seen: (1) high variability of dances in the hive, (2) buzzing flights at a feeder, and (3) the importance of flower scent adhering to the bodies of foragers. Taken together, these provide the view (Figure 13) that the inaccuracy of the dance is combined with additional active recruiting in the field.

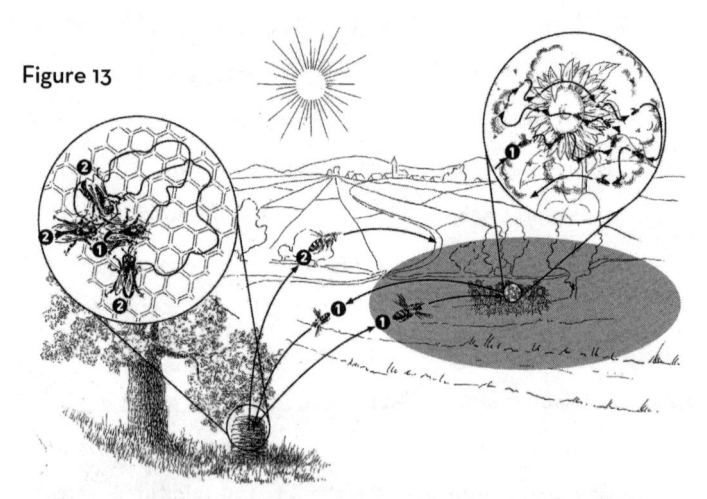

Figure 13

Recruits (2) that followed a dancer (1) begin their foraging flight in a 'corridor of uncertainty' (grey area). Field bees (also dancers [1]) flying back and forth between the hive and the feeder help recruits to find the goal by scenting during their buzzing flights and adding to the attraction of scent from flowers (from Tautz 2015).

The dance alone sends recruits off in an approximate direction[27] and provides them with an estimate of the distance to the goal.[28] The distance estimate becomes less exact the more remote the goal. However, recruits do not reach the goal only through the help of the dance that sent them off and by following dance instructions. They also follow experienced bees, lured by the pheromone they spread at the site during buzzing flights, and by the scent of the flowers. When 'sending' and 'luring' fuse seamlessly together, recruited bees certainly reach their goal. Incongruity between the sending and luring initiates a search phase in the area of the indicated goal. Searching bees attempt to discover additional goal-finding cues. If this is unsuccessful, they return to the hive empty-handed. The bee dance is therefore only part of a complex communication comprised of sending and luring; a linked sequence of behaviour in the hive and in the field.

A different view and its consequences

The original model of Karl von Frisch (see Figure 8 – model 1 in the following) and the somewhat more complex model just described (Figure 13 – referred to as model 2 below) lead to considerable differences in the perceptions of bees' abilities and what is expected of them.

Figure 14A

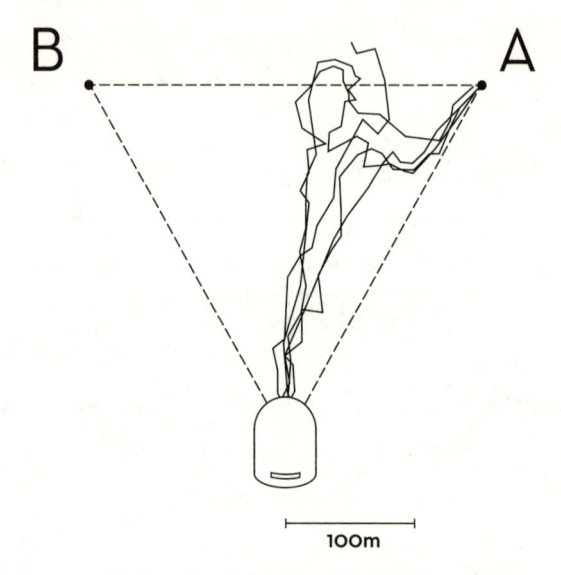

100m

The radar-recorded paths of the first flights of foragers recruited by dancers in a hive to a feeder (Menzel et al. 2011). Bees fan out from the hive in a specific direction and then enter a search phase. The last stretch of the flight, over nearly 100 metres, precisely follows the shortest route from the hive to the feeder at A. The surprising sudden turn along the correct route to the feeder could be explained by a meeting between inexperienced and experienced foragers (Tautz & Sandeman 2002).

The following example in which flight paths of marked foragers were tracked with radar and recorded clarifies this.[29] The

Figure 14B

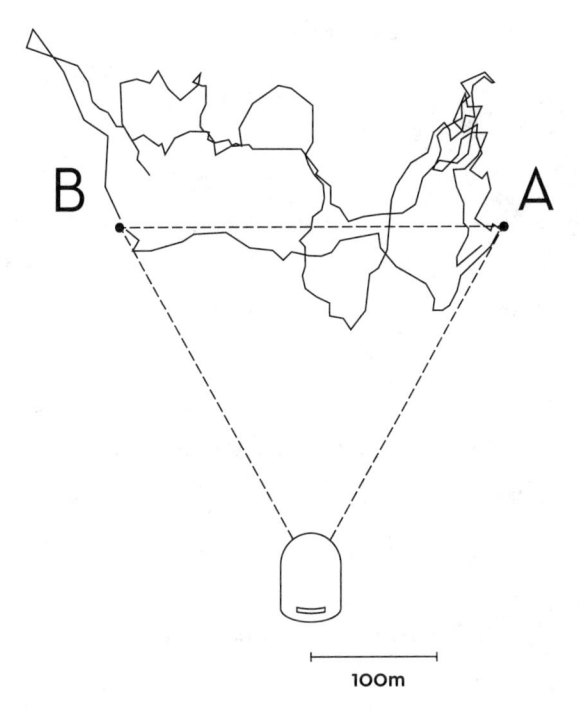

If no food is offered at feeder A, some bees do not return to the hive but fly to B (three traces extending from right to left), where a second feeder, previously visited by bees, had been dismantled. The flight paths from A begin along a very broad path and meander (a typical search pattern of insects following an odour trail) before ending at B (from Menzel et al. 2011).

flight paths of recruited bees that flew from the hive to a goal at point A can be seen in Figure 14A. Figure 14B shows how bees flew from point A to a second goal at point B. The background to these flight paths is the following experiment: A feeder was placed at point A and another at point B and neither were artificially scented. Thirty foragers were trained from the hive to feeder A. The flight paths of the recruits that followed the dances for A were recorded (Figure 14A). In addition to the bees trained to feeder A, up to two bees were trained to feeder B. Reducing the sugar concentration in both feeders stopped the foragers from dancing in the hive and therefore no recruits visited the feeders.

On the following day, no food was offered at A; instead, the supply at B was made so attractive that both bees trained to B danced in the hive. Recruits that had flown to A now witnessed the dances for feeder B in the hive. However, they did not fly to B but to A, a location they already knew. During the course of the observation in the field it was noticed that several bees from group A did not return to the hive when they found nothing at A. Instead they visited B, a place where they had not been before, but which the two bees trained to B had just advertised in their dances. This occurs only if A and B are close enough together. Doubling the distance between the hive, A and B, resulted in no further direct flights from A to B.

The interesting question raised here is how the bees that were recruited to A, and having never visited B, found B without first flying back to the hive.

Basing the results on model 1 (precise directional information from the dance alone) one must invoke some complex assumptions in order to explain the behaviour of the bees. According to model 1, bees that received no reward at A but witnessed dances for B, used the information from these dances to re-orient themselves. They can do this if we assume (1) they have exact information from the dance, (2) they can determine the vectors between relevant points, and/or (3) they have access to a cognitive map, that is, a map in their heads.[30] Disappointed bees at A get out their navigational aids and plot a course to B as their new goal.

Must one assume that bees flying between A and B, without returning to the hive after finding no food at A, execute a complex intellectual task? Why so complicated when it could be so much simpler? Model 2, combining the motivation in the hive to fly (the dance) with attraction in the field (scent) can explain with far fewer assumptions why the disappointed bees at A flew to B. Nature has provided honey bees with an extremely sensitive olfactory system that expresses itself in the detection of the scent of flowers and chemical signals between bees. An alternative to invoking complex navigational calculations by foragers, is as follows: Bees finding nothing at A proceed as usual when they arrive within a goal area and cannot immediately find the nectar source. They circle and 'sniff around' to see if there are any additional clues to a profitable food source. In this way, they discover the feeder at B, where, because it has been turned into a rich source, has also been scented by arriving bees performing buzzing flights. Could it be that simple? Nothing works in the bee world without scent. If we reject model 1 and admit that the behaviour of the bees is adequately explained by model 2, we have still reason enough to be amazed at the abilities of honey bees.

The drones: Callboys for the queen

We have now heard at length about the workers, their various assignments and their many talents. Occasionally we also mentioned the queen and said we would get to know her better later. First, though, a little about the lords of creation, for men do exist in the bee world, and they fulfil practically every cliché applied to males of any species. Drones (male bees) represent the insect version of the 'typical male'. They have enormous eyes with which they see very well. And what do they look at? Girls, naturally. When not tripping around, they hang out in the hive, get in the way or allow themselves to be taken care of. They have only one thought in their heads: yes, of course, what else?

Only half a man

In fact, drones are actually half female. Before the queen lays an egg, she explores the size of the comb cell with her antennae and head. If the cell diameter is small, she lays an egg that has already been provided with sperm from her store, and the fertilised egg develops into a female worker. Comb cells with large diameters receive unfertilised eggs. Drones develop from these unfertilised eggs, which genetically are half-creatures in comparison to the female workers. Drones possess only the queen's set of chromosomes, which are contained in the egg, and no additional chromosomes from a male sperm cell.

Johann Dzierzon, the Catholic priest who first described this phenomenon in the nineteenth century, had a problem with his employers. The Church could not accept that any being other than the Virgin Mary, Queen of Heaven and mother of Jesus, could bear male heirs without the participation of another male. But after considerable discussion, Dzierzon was finally awarded an honorary doctorate. After all, the queen bee does not bring holy redeemers into the world, only male bees.

Some time is also needed before the genetically simple drone develops from the egg. Queens hatch after sixteen days, workers after twenty-one days and drones after twenty-four days. They are slow in this regard too.

And lazy. Emerging workers immediately begin to clean the cell from which they emerged and then join the cleaner bee service. A young drone at first does nothing at all, and then goes begging. Workers feed the young drone in his first few days. Once he has developed a little strength, he helps himself to provisions. He lounges about the hive, sips the nectar and consumes a lot of pollen. The protein-rich diet is important to prepare him for his assignment. Drones are the callboys of the hive. Their task is to fertilise willing young queens. Eight to ten days after emerging, the callow youth develops into a mature man. He is now sexually mature and carries about 10 million sperm cells in seminal glands in his large abdomen.

Such potential and fecundity is unsettling. Although as a young drone he may have made a few orientation flights around the hive, now is the time to begin the serious gadding about, and he is not alone. Not only buddies from his own hive, but also those from all the other hives in the area get underway looking for brides. They meet in so-called drone congregation areas. It is not known how these arise. They occur in specific regions where, year after year, large numbers of drones can be observed in the summer months. Drones in these areas circle around at an altitude of between 10 and 20 metres, waiting for young receptive queens from neighbouring hives to appear.

Drones are well equipped to find queens. Their enormous eyes cover virtually their entire faces. They also have an excellent olfactory sense to detect the scent trail of a queen and possess tufts of bristles on their hind legs to catch and hold onto a queen while flying.

Sperm bomber attacks

The detection of a receptive young queen leads to a wild chase, because Her Majesty is usually spotted at the same time by many large-eyed drones. A swarm of drones fly after her, each attempting to fasten onto her. But it is literally the end for the winning drone. In order to mate successfully, the drone must turn his genital organs, which lie deep in his abdomen, outwards and attach these to the queen. For him, this is a fatal process. He dies after ejaculation and falls to the ground, leaving his ripped-out genitalia anchored in the queen. She carries with her this 'fertilisation mark', as beekeepers call it, which workers remove when she returns to her hive.

The great majority of drones do not find a queen partner and die when they are thirty to forty days old. Exhausted from continual searching and flying about, the day comes when they no longer return to the hive – at least those that were born early in the year. A less pleasant fate awaits those still engaged in their loose lives when the period during which drones are tolerated by workers comes to an end. The drones' task is to fertilise queens, and unfertilised queens are present only during the swarming period – from the end of April until about the middle of July. The colony does not need drones when no young queens are expected. The 'drone slaughter' begins about the middle of July, the term suggesting that on a fine day, battle-hardened workers close ranks, turn on the previously nurtured men and kill them off. Sometimes, after July, one can in fact see a worker attacking a stingless and defenceless drone at the entry to the hive, but there is no general massacre. Colonies that no longer intend to swarm raise less drones. The court of workers that accompany the queen during egg laying see to it that she lays very few eggs in drone cells, and so there are fewer male offspring.

Drones that are still in the colony when it is decided that they are no longer needed seldom die from being stung. Instead, the workers allow them to starve to death. Begging is no longer rewarded and they are driven out of food storage combs and out through the hive entrance. Guard bees prevent them from re-entering the hive. The time of drones ends at the latest in October and very rarely can one find a clever drone in a winter cluster that has somehow managed to win the right to stay. Even so, although drones are only seasonal helpers in the Amazon state, without them there would be no more honey bees.

Fatherless companions

Sexual reproduction usually rests principally on the combination of two different types of germ cells: male sperm and female ova fuse to produce a new organism. Both participating germ cells bring a complex set of chromosomes, which store genetic information, into the union. The random new combination of genetic material results in mixing and diversity, and new traits are possible in newly produced individuals. This is known as bisexual reproduction and it is a deciding factor in evolution.

Because this is the case, one can assume that every individual of a bisexual species will have a father and a mother. There are also two sexes in bees, namely the female queen and workers, and the male drones. Nevertheless, drones develop from unfertilised eggs. These contain only the sixteen chromosomes of the queen. The only male genes drones receive are those carried on the chromosomes of the queen, who developed from a fertilised egg. Therefore, genetically, drones have no fathers, only grandfathers, while female bees have equal shares of their father's and mother's genes. What are the consequences of this for genetic diversity in a bee colony? Do bees forego the opportunities offered by bisexual reproduction for the evolution of the species? Here, we need to take a closer look at the admittedly somewhat confusing world of chromosome transmission.

Cells containing a single set of chromosomes are called haploid, those with a double set are diploid, and those with more than two, polyploid. Workers and queens in a bee colony are diploid, with two sets of chromosomes present in their body cells. Drones are haploid, because they have one set of chromosomes. However, this is true only for the drone egg and larvae. The egg cells possess sixteen individual chromosomes and are haploid, and the embryo within the egg also has only sixteen chromosomes. With hatching and the beginning of larval development, genes in the cells of drone larvae multiply. Only nerve cells and germ cells remain haploid. Up to ten copies of the original sixteen-chromosome set can appear in other body cells of a drone.[31] Therefore, in terms of chromosome numbers, drones are not haploid, but they are haploid in respect to their genetic heredity. Their sperm cells still have only a single set of the chromosomes from their mother, and there is no paternal set. Drones cannot, therefore, pass on different genetic combinations. Species with diploid males can, because during the maturation division during sperm production (meiosis), chromosomes are randomly distributed. Haploid drones have nothing to distribute, because all of the sperm cells are identical. The crucial advantage of bisexual reproduction – the random separation and recombination of chromosomes and the production of novel combinations – appears to have been lost by bees. But only at first sight.

The coquetry of young queens and the promiscuity of mature males compensate for the genetic naivety of drones. This takes place at drone congregations. Drones from many colonies assemble with the expectation of meeting virgin queens, and multiple pairings of queens with drones from different colonies ensures that a multicultural genetic diversity dominates in the colony. A laying queen has access to sperm cells from many different males, and so although all bees in a colony have the same mother, female offspring of the queen have different fathers. Every bee colony has as many full sister lines as drones that paired with the queen. Overall, a bee colony consists of a large collection of half-sisters.

Beekeepers who place no importance on race purity see this diversity especially well. Three races of bees are found in Europe: the Carniolan honey bee, the Buckfast bee and the European dark bee. The European dark bee was originally indigenous to various ecotypes north of the Alps following the last ice age. Apart from a few surviving stocks, they virtually disappeared from western Europe, although recently they have attracted the interest of hobby apiarists. Carniolan honey bees, originally indigenous to south of the Alps, particularly Carinthia in southern Austria, were introduced to large areas of northern Europe during the middle of the twentieth century. They are considered to be easier to keep and produce more honey than the European dark bees. Buckfast bees are a cultured race, developed in the twentieth century by Karl Kehrle, also known as Brother Adam, at the Buckfast Abbey in England. They are particularly well adapted to a cool and changeable climate.

If drones from various races enter drone assemblies, workers can be discovered in colonies that have the slightly grey colouring of the typical Carniolan bee, while their half-sisters are browner and exhibit the wide yellow abdominal rings of the Buckfast bee.

Where apiarists also keep European dark bees, these turn up in the colonies, although the queen is a carnica majesty. Considering that one queen can produce all these genetically different workers and new queens, then it is clear how genetic diversity is maintained in a colony despite the genetic simplicity of the drones. Apiarists confirm that the most vital colonies are not the race-pure, but those where the queen had the opportunity to pair with a variety of different male types and produce the most diverse and fittest offspring.

Epigenetics or why bees 'bake' their sisters

The combination of genes in a fertilised egg cell determines the characteristics of the resulting organism. But it is wrong to assume that the genetic disposition of a fertilised egg initiates an inflexible program of development ending in a single, very specific, result. The progress of development is also determined by epigenetic factors imposed on the initial genetic characteristics. An example of an epigenetic factor in a bee colony is the diet supplied to larvae. An egg laid by the queen with basic genetic characteristics can become a worker or, if the larva is fed royal jelly, a queen. There are also more subtle examples.

Bees can alter the nature of their sisters in other ways. What sounds like science fiction is based in the brood nest, where workers control development of their sisters from egg to their emergence from the pupal cell. Feeding can influence larval stage, temperature and pupation. Temperature regulation is a control that can be adjusted with great sensitivity and employed to manipulate traits in the next generation of bees. The long-lived winter bees, for example, develop at brood-nest temperatures that are lower than normal, and heater bees set this temperature. They 'bake' their sisters in a slow oven, allowing them to reach a much greater age than they will themselves.

Bees can apparently also pre-set the willingness and effort workers expend in carrying out certain activities through regulation of the brood nest temperature. Ventilator bees that reduce the temperature of the hive by standing at the entrance and fanning air into the hive with their wings are a good example. Natural temperature changes occur gradually, so if ventilator bees begin cooling at the first sign of a temperature rise, a small increase would quickly be countered. However, if all ventilator bees set to work together in response to a small temperature increase this would result in a sudden and significant fall in temperature, which would then have to be compensated by heating. A waste of time and energy. Instead, the behavioural response is appropriate to the need. Only if the first few ventilator bees are unable to lower the temperature do additional helpers join in. If the expanded group still cannot cope, then even more ventilator bees must contribute to the activity. Should the sun's rays lead to an increased temperature in the hive, more bees appear on the landing stage and begin fanning, resulting in the comb passage temperature remaining constant. How is this graded response of ventilator bees managed?

Bees carry out particular tasks in appropriate places and times, when signals in the environment release these behaviours. But bees are not all equally sensitive to such signals. For ventilator bees, only a few of them respond to a small rise in temperature, but more of them are motivated by a stronger stimulus.[32] So it is that bee colonies adapt their actual activity to current challenges purely through the different sensitivities of colony members and without a need for central control. The supposition is that the range and gradation of sensitivity thresholds in a colony increases with genetic diversity influenced by epigenetic factors.

Figure 15

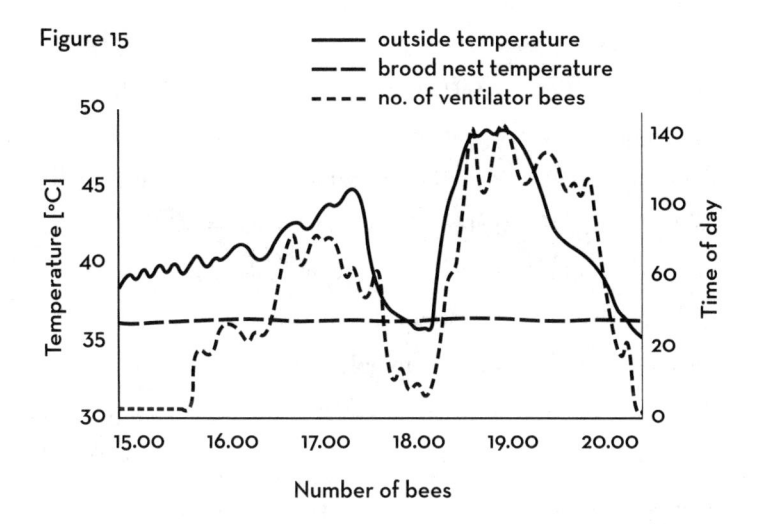

— outside temperature
— — brood nest temperature
- - - - no. of ventilator bees

Despite extreme variation of outside temperature (the temperature of the hive walls in the sun), bees manage to hold the temperature within the brood nest constant. The secret lies in the employment of an appropriate number of ventilating bees to renew the air in the hive. The dotted line shows that the number of activated ventilator bees closely follows the temperature that has to be regulated (data from HOBOS, 30 June 2012).

The queen bee: A monarch with limited power

'*L'etat c'est moi!*' Louis XIV, ruler of France in the seventeenth century is supposed to have described his role as king with these words: 'I am the state'. In fact, Louis XIV probably never said this, but the statement is not far from the truth. At that time, the king of France stood above the law and his rule was absolute. He was not dependent on, nor responsible to anyone for his decisions. What he decreed was carried out. Everything in the king's court in Versailles rotated around the monarch. The polished court ceremony of this time is legendary and the monarch so totally the centre of attraction that no-one at court – and all who wished to be someone in France were there – could escape the expectation and requirements of the protocols. An honour could be bestowed on a baron by choosing him to hold the king's shirt sleeve when the king dressed each morning.

And the queen in the bee colony about whom we have heard so much, is she an absolute ruler? Does she decide the fate of her people arbitrarily, and is she the centre of attention in the small kingdom in a box? Yes, and no.

The ruler's perfume

A queen bee is unique, for every colony normally has only one queen. Even if in summer there are 50,000 individuals in the colony, of which a few hundred are drones, there is only one queen, and she is appropriately recognised by the members of her colony. Nurse bees accompany her continuously, feed and clean her, and provide her with all she needs. Like all monarchs, she has a court. She also exercises considerable influence over her colony. And she has power. Not through laws, administrative rules or the military. Her power is her perfume.

A queen produces a 'queen substance' from her mandibular glands. This mixture of pheromones is most important for 'hive harmony', as apiarists say. Like hormones initiating metabolic processes in our bodies, or by their absence causing these to fail, so pheromones can release specific behavioural reactions in individuals of the same species. Pheromones that serve as sexual stimulants are perhaps the most obvious. A female cat in season exudes an odour that induces male cats in the area to howl with desperate desire. The stimulating odour of a bull elk's urine leads elk cows to take an aphrodisiac bath in it.

Similarly, the queen's perfume ensures that everything runs according to plan in the honey factory. It keeps everything going. The queen's perfume is distributed by court bees and through the habit of bees grooming and feeding one another. Court bees absorb the perfume from the queen and pass it on simply by coming into contact with other bees in the hive. These in turn groom other bees and so transfer the perfume to them.

If the queen substance is absent, perhaps because the queen has died, the colony is disturbed and behaves abnormally. Experienced beekeepers notice this behavioural change as soon as they open a hive and begin to withdraw combs. Colonies with queens at first buzz out but then soon settle again. Very few workers are upset and attack. Colonies without queens also buzz out but do not settle. Instead, more and more join the buzzing swarm until almost all the workers are out, covering the combs and making a loud noise. If beekeepers find no uncapped brood cells or cells containing eggs in a hive that makes this particular sound, they know the colony is 'queenless'.

Changes in the executive suite: How bees replace their CEO

The queen exercises her power through her perfume, but if it fades, this can be used against her. The queen is not an absolute ruler, but the colony is still dependent on her. However, if she is no longer able to perform her duties well enough or at all, she is replaced – and not necessarily by the apiarist, but by the bee colony itself. Workers in the honey factory get themselves a new chief. How does this happen?

Suppose the queen of a colony has served her hive well for four years, laying more than a million eggs from which workers and drones emerged. Now she is old and the colony is aware of this. In the fifth year, there are fewer larvae to feed than there should be. Many of the eggs that should have produced workers were unfertilised because the sperm reserves the queen obtained during her nuptial flight are depleting and the unfertilised eggs are eaten by the nurse bees. The brood nest, usually an almost completely closed area of worker cells, has many empty cells. In addition, the radiance of the queen is diminished; she cannot produce enough queen substance to keep the concentration of her perfume high enough in the hive to satisfy her colony. They notice. Their queen is no longer able to cope.

In such a situation, workers build a queen cell, using their mouthparts to scrape away empty cells in a brood comb over an area of about 2 by 3 cm, to make a shallow hollow. A small wax bowl, about 1 cm long, is constructed at the upper end of the hollow. Nurse bees lead the old queen to this bowl and she lays the egg from which her replacement will develop.

This egg is not, as one may expect, something special. Under normal circumstances it could hatch into a worker bee. But nurse bees notice that the egg and the larva are in an unusual cell. Neither a worker nor a drone will arise here, for the cell is larger, rounder and longer and projects out from the comb. A queen will be born here. The larva in this cell will not be fed honey and pollen. This larva gets royal jelly and nothing else. It is literally immersed in a thick layer of the whitish pudding.

This pudding is the queen's secret power. The queen larva's diet of royal jelly leads to the activation of different genetic switches than those activated in the development of a normal worker bee. Sixteen days later, the normal egg laid in the queen cell is a young queen.

The colony now has two queens: the older still laying eggs, and the crown princess. The princess is not yet in a position to take the place of her mother, because she is not yet mature, nor has she mated.

The new queen becomes sexually mature five to six days after emergence, when she has begun to produce a clearly detectable amount of pheromone. This is a signal to the colony: 'I am here.' And also a signal to the old ruler: 'Reckon with me.' The young queen now has a small court protecting her and seeing to it that nothing untoward happens to her. The old queen could come too close. Who knows, perhaps she realises that her star is sinking. Queens can sting and kill competitors. The young queen is now in heat and wants to visit the boys.

Workers grant her wish around noon on a fine day. A 'play flight' swarm forms in front of the hive entrance. You will recall that a number of bees occasionally gather in front of the hive and fly back and forth. Beekeepers first thought this phenomenon was due to the practice flights of young bees learning their surroundings before becoming foragers. Young foragers are present from April to October, but play flight swarms appear from May to the middle of July. Very few young bees take part in play flights; instead, the majority are experienced foragers and it is these who are important here. The small swarm in front of the hive waits to receive its young queen, who is ready to fly out of the hive on a nuptial flight. They fly off with her in their midst to a drone congregation. The queen has never been out of the hive, let alone flown any distance from it and is totally dependent on the accompanying swarm. She could not find her way home if she were to lose her escorts, and would die. Her perfume, which will attract the attention of drones, ensures that her escorts always know where she is and they stay nearby when she pairs with a drone. Her Majesty is not particularly discriminating, and certainly not satisfied by pairing with only a single male. She can cause up to fifteen males to lose their minds and lives, and to do so some queens will visit the drone congregation on several consecutive days.

Sperm stores in her abdomen are then filled and she is now a mature queen ready to lay eggs and ensure the survival of her colony for the next four to five years by carrying out a single task – laying eggs. This is her only mission. She is tended around the clock, and provided with all she needs so that she can perform her duty optimally. She continues to be fed royal jelly for the rest of her life, because only with this high-quality nourishment can she lay the enormous number of eggs expected from her.

Mated queens begin egg-laying immediately after concluding their nuptial flights and the colony soon has a fine brood nest with many capped cells. The queen's perfume is strong and the colony knows that all is well. The queen is there.

And the old queen? Some beekeepers claim to have had colonies in which they saw two queens. It is possible that the old queen is tolerated for a short time during the presence of the new queen. But this will not last long and eventually the old queen disappears. Perhaps she is killed by the young queen if she comes too close to her, or perhaps by workers now imprinted with the perfume of the new queen and who no longer recognise the old queen as a colony member. The old queen may starve because her court deserts her when they detect a new perfume drifting through the hive. A queen bee maintains her power only as long as she is able to serve her colony. Exhausted, she ends her time according to the will of her subjects.

The 'Bien': An intelligent superorganism

Workers with their many different tasks, drones and the queen are members of a team that together guarantees the survival of the colony over the seasonal rhythm of years. This community of many individuals solves problems by work-sharing and communicative exchanges that single individuals could not solve on their own. A queen bee could not produce offspring alone; a single worker could not build a comb; and without the community of workers in the hive, drones would be helpless and their existence pointless. A community of individuals, dependent upon one another, whose achievements cannot be attained by single individuals but instead are based on communication and cooperation is, from a biological perspective, a superorganism. A superorganism can be thought of as a being with a single body, except that this body does not consist of single cells with various functions interacting in different ways, but of separate individuals. The 'bee colony' superorganism can be perceived to be a single, breathing, living entity and is sometimes referred to as the 'Bien'.

Bee intelligence: How bees see the world

Observing life in a bee colony and its complex integration of many different assignments and procedures, it is difficult not to assume that it's all centrally planned and managed.

But who in the bee colony is acting in an intelligent and systematic way? Single bees that decide for themselves? Or is intelligence in a bee colony located not in individuals but in the sum of these individuals – that is, the entire colony? Has the bee colony developed control structures in which single bees, like cells in a body, recognise their particular assignments at particular times?

Such questions are the object of intense study in the field of sociobiology, a special area of behavioural biology. A highly complex set of concepts has arisen from simple basic assumptions relating to how bees 'decide' or how 'intelligent' they are. Despite many theories concerning these questions, their basis is always the sensory perception of single bees. When these or the entire colony are to arrive at 'intelligent' decisions, they need information about conditions and events in their surroundings.

Bees determine these with their sensory organs. We know a great deal about the olfactory and visual senses of bees. These are their 'windows on the world' and the foundation for their ability to find their way among the flowers.[33]

What bees see: A grey-coloured world in slow motion

To see the world as a bee does would be a catastrophe for us. We could forget about reading newspapers, driving cars, going to football matches, or using laptops and smartphones. A bee's world is very different from ours. Their vision is constructed to solve the unique problems facing a colony that must survive and multiply.

Unlike us, bees have two visual fields, one for the right eye and one for the left, and a broad area of blindness between them. The images a bee receives when sitting still are grossly pixelated, for each eye is made up of 6000 small, single eyes, called ommatidia. Each ommatidium sees only a very small part of the world. We can get an impression of what a bee sees when it is not moving if we look at the face of someone through a bundle of drinking straws: a patterned, structureless something, with no contours.

This image changes as soon as the bee moves and single images in each ommatidium fuse together. We can experience the same phenomenon if we look through a window with a fly screen. A vehicle number plate, for example, seen through a fly screen, appears pixelated if we do not move our heads. But if we turn our heads, the grid structure disappears and we see the number plate clearly.

This effect is due to us slightly increasing the number of images we perceive by turning our heads. Bees have not two but 12,000 eyes. They need considerably greater image motion to perceive movement. What appears to a bee as a single image 'slide show' fuses into a movie only when these images arrive at the eye at a frequency of 250 per second. This also means that bees do not see fast movements as blurred, but instead in 'slow motion', and they are able to recognise one another during flying. This is important when swarming, when accompanying a queen and following her at a drone assembly, and when buzzing around a feeder in order to attract foragers to the best nectar sources. Knowing that bees are better able to see rapid, rather than slow, movements is good for beekeepers. Those who wish to look inside a hive or work with bees are advised to avoid hasty actions. Moving a hand quickly over an open hive is soon seen by bees, who fly towards it with stings ready. In contrast, a slow hand remains undetected, and practised apiarists seldom wear gloves when working with their bees. Anyone who slaps about in panic when a bee flies around them provides a perfect target.

The resolution of images received by bees changes with their flight speed and their perception of colour is dependent on how fast they are flying. Bees see colour but their impression is completely different from ours. Only three basic colours exist for bees – green, blue and ultraviolet – and all their impressions of colour are gained from combinations within this spectrum. The capacity to see the colour green is an ancient ability for insects, developed even before flowers existed, when the dominant colour was the leaf-green of chlorophyll. The first insects needed only to distinguish between this, the mineral colours of the earth and the blue sky. Perception of the polarised portion of ultraviolet light from the sky, produced when it passes through the atmosphere,[34] provides cues for orientation. Bees see both short-wavelength ultraviolet light that we cannot, and also the plane of its polarisation, allowing them to locate the sun's position even on cloudy days. In addition, their sensitivity to ultraviolet light helps them in their visits to flowers. Many flowers have areas on their blooms that reflect ultraviolet light particularly well. We are unable to see this, but for bees it is like having landing lights on an airfield.

Like ourselves, whether bees see colours or not depends on prevailing conditions. We only see colours in bright light and lose this ability as the light decreases in intensity. All cats are grey at night! Bees lose their colour vision even in bright daylight when they fly quickly, and they can reach up to 30 kilometres an hour. Colours fade for them at flight speeds of only 5 kilometres an hour,[35] above which only those visual cells sensitive to green are active. Bees then see the world as though through a green filter, and everything in different shades of green. Objects that reflect no green wavelengths appear black to them; those that reflect only green, appear white.

This curious disappearance of colour is practical for bees. The overall structure of the landscape and the detection of obstacles is important for them, and their aims are to complete a flight without losing orientation and to avoid obstacles. Coloured flowers appear only when they reduce their speed and prepare to land. But the more colourful the world becomes, the less sharply it is defined. Bees have developed an additional strategy to steer their way to the flowering goal – their sense of smell.

A cosmos of perfume and tension: How bees detect odours and what they can smell

An olfactory sense is of critical importance for all flying insects searching for a goal. With this sense, dung beetles find fresh cowpats, wasps find tasty plum tarts, and bees find flowers offering pollen and nectar. The perfume of flowers, which evolved alongside nectar-drinking insects, plays a double role for bees.

First it helps them to discriminate between different plant species. This is important when it comes to efficient collection from a profitable nectar source. Scouts must accurately inform foragers of what is available, so they bring a taste and perfume sample of their find with them into the hive. Put in the picture, foragers can then fly out in the directional 'funnel' indicated to them in the scout's dance. Within this funnel – and this is the second function of flower perfume – they discover where to find the nectar source from the scent of a particular flower and pheromones released by foragers already at the site.

An olfactory sense is not only important for finding profitable nectar sources. Drones at congregations find queens with its help. A queen's perfume holds the entire colony together. Bees that are attacked can call their sisters to their aid with the odour of their venom, alerting them to danger and its location. Bees determine the source of an odour with their olfactory systems the way we are able to tell where a sound is coming from with our ears.

A bee's olfactory sense consists of about 10,000 sensory receptor cells carried on their antennae. This pair of independently movable antennae provides them with the ability that we, with our single, fixed noses do not enjoy: namely, bees can detect odours spatially – that is, they have a three-dimensional perception of an odour field.[36] We find it hard to imagine what it would be like to perceive the three-dimensional shape of an odour cloud and locate its source. And our imagination fails us completely when we consider another sense that we do not possess at all: that of being able to detect electrical and magnetic fields.

We have known for more than fifty years that the earth's magnetic field influences the way honey bees construct their combs and that it can also disturb their dances in the dark hive.[37] For forty years we have known that the cuticle (outer covering) of bees can accept and hold an electrical charge.[38] Highly sensitive instruments reveal that flowers express a particular pattern of electrical charges strong enough for bees to detect.[39] Honey bees experience both very weak and strong electrical and magnetic fields. They are 'under tension', so to speak, which at times can be a problem for them.

Now and again the sun emits large streams of particles into space, in a solar 'wind'. When these reach earth, they can severely distort our magnetic field. During particularly intense emissions from the sun, bees appear to be unable to find their way back to the hive. The earth's magnetic field could provide bees with a sort of compass like that used by migrating birds. When this is disturbed by the sun's emissions, it no longer provides the correct information. Bees that have not already 'flown in' to their hives may now have the wrong 'map' and so cannot find their way.[40]

Learning, planning, discriminating: The criteria of intelligence in bees

One can be mistaken only if one is able to choose between two alternatives. In order to recognise possible alternatives, individuals must be able to effectively process information from their environment, store this and recall it. We must be able to learn. The honey bee's ability to learn has continued to surprise researchers since they began to study its nature. Bees learn the location of their hive after a few exploratory flights; they recall colours and forms after very few training sessions; they remember the time of day that food is present at a particular place;[41] and can associate a food reward coupled with a perfume after a single trial. It does indeed appear that bees can act according to a plan.

Flowering plants have developed a wide variety of forms, colour and perfume in their effort to attract pollinating insects. Different plant species also open and close at specific times of the day and their production of nectar can also follow a similar temporal biorhythm. The Swedish naturalist Carl Linnaeus first explored this fascinating phenomenon in the eighteenth century so thoroughly that one could construct a flower clock from the daily flowering times of different species, and this informative and interesting application of a fundamental natural observation can be found today in many parks and gardens.

For pollinating insects such as honey bees, this chronobiological phenomenon means that different nectar sources are accessible at different times of the day. The situation for foraging bees becomes more complicated, because different flower species usually grow in different localities. Can bees under such circumstances organise their day by learning and planning which plants to visit and when?

That bees have such a capacity was demonstrated by Australian researcher Shaowu Zhang and his bee research group in 2006[42] and 2007.[43] Free-flying bees were trained to enter a system of passages to find food. The single entrance to the passages led on to a Y-shaped intersection, presenting bees with a choice of two alternatives, each of which was marked with a specific optical pattern. Entering bees received a food reward only if they chose the correct optical pattern, which they quickly learnt. Bees also learnt to adjust their behaviour if the patterns changed at different times of the day. They knew that a pattern promising food in the morning did not mean food in the afternoon, and so chose the alternative, whereas the pattern that offered nothing in the morning was a sure sign of a positive reward in the afternoon.

The ability of bees to remember the association between the time and place of a reward (the correct passage) and the nature of the reward (nectar) is referred to by psychologists studying learning in humans as 'episodic memory' and regarded to represent a type of long-term memory. The achievement of bees satisfies the criteria of episodic memory and has therefore been termed 'circadian timed episodic-like memory'.[44]

Bees not only learn and remember. Researchers at the University of Würzburg have shown that bees recognise and distinguish between up to four different objects.

If we are shown a box containing four objects we are usually able to quickly appreciate what is in the box and recall this when the box is taken away. Should the box contain five or more objects, we need time to count them in order to remember correctly. Apes, pigeons and other vertebrates are also able to recognise only four objects or fewer at a single glance. Honey bees are the same.[45]

This can be shown by setting up two boards next to one another, one of which displays an image of a single object, the other of two objects. Both boards have a hole through which bees can fly. Bees receive a reward of sugar water only when they fly through the hole marked with two objects and they quickly learn this. Approaching bees are not confused by changing the colour or shape of the forms. They fly only to the board with two images. They are unconcerned by the right or left location of the boards, or whether the images are red apples or yellow circles. There just has to be two images, because to them that means food.

Bees trained to boards with two and three, or three and four, images always quickly learn where they have to fly. They fail only when they are confronted with more than four objects.

Apparently bees can correctly estimate small numbers of objects. They may use this skill to gauge the number of flowers on a branch or bees on a flower, and decide between the options of landing or taking off.

A variation of the above experiment led to a further surprising realisation. Bees recognise 'pictographs' or structural analogies just as we do. Those of us interested in art quickly learn to recognise the style of a picture and can assign it to a particular artist even if we had not seen that particular painting before. A Picasso differs from a Monet in respect to some typical characteristics.

It occurred to a Brazilian–Australian team of researchers led by biologist Judith Reinhard to explore bees' abilities in relation to this task. Bees were presented with two boards, one of which displayed an impressionist painting by Monet, the other a cubist Picasso. Food was presented at only one of these. Once the bees had learnt which artist was associated with a reward, they were exposed to an exhibition of works from both artists, none of which they had seen before. A reward was provided only if the bees made the correct choice between the two artists and, although less certain in their choices, they achieved a success rate that was greater than pure chance. Bees can apparently extract characteristics from complex structures, categorise and remember these.[46]

Traditional companies

Individual bees as members of the bee colony superorganism are remarkable performers. But they do not possess all these talents from the moment they crawl out of their cells. Instead, they have the potential to acquire these skills, but can develop them only within the colony community. The conditions for transfer of knowledge lie with the unusual learning ability of bees and uninterrupted contact with other colony members. Fifty years ago, Martin Lindauer, a very well-known behavioural biologist and bee researcher, showed in a series of experiments that bee colonies can pass knowledge on as a 'tradition'.

Lindauer trained foragers of a colony to visit a feeder open between 5 and 6 am, an unusual time for bees. The foragers soon learnt to only visit the site within these times. The colony contained closed pupal cells in its brood nest, normal for summer.

A second colony was trained to a second feeder that remained open all day. Lindauer now took closed brood combs from the first colony and allowed them to emerge from their cells within the second colony that had a feeder open all day.

TEAMWORK IN THE HONEY FACTORY

The question behind the experiment was to see when the young bees from the first hive foraged, after emerging in the second hive. Would they accompany the foragers of their new colony on their first flights and collect all day? Or had they learnt, while still in their pupal cells, the time window during which their old colony foraged? Foragers in the old colony collected between 5 and 6 am. Bees transferred as pupae to the new colony did exactly the same. This also applied to pupae transferred from colonies trained to feeders open at between 8 and 9 pm. On emerging, these bees also foraged at this late hour.[47]

Behavioural performance is normally based on the properties, abilities and activity of the nervous system, and in particular the brain. Is it possible to identify something that reflects this in the brain of the socially highly organised and unusually talented honey bee, in comparison to solitary carrion-seeking flies?

So far, the answer is no. It does not seem to be possible to determine from examining their brains whether an insect comes from a socially organised and labour-sharing superorganism or is alone. However, this is possible if we examine the digestive system. We have already heard that worker bees nourish the queen, satisfy the hunger of the larvae and drones, and also feed each other. Trophallaxis, the exchange of food between workers, encompasses much more than making sure that all in the hive have enough to eat. Trophallaxis is the epitome of communication methods among social insects. Bees experience the situation prevailing in the hive through the exchange of food and thereby also odours. What is the nature of the nectar being brought in? How much pollen do we have? Is all in order? Do we have a queen?

Harmony, cooperation and the exchange of information are guaranteed because bees are essentially compañeras, comrades who share their bread. The anatomy of the honey bee gut provides good evidence of this. They have a 'social stomach', the 'honey stomach', which is a highly expandable portion of their gut located in their abdomens just before the digestive tract. Collected nectar is held in the honey stomach and brought back to the hive, regurgitated there and processed into honey. A very small portion of the nectar in the honey stomach of a forager is allowed to move on into the digestive tract. Honey stomachs can also be filled in the hive and then distributed to others. Collaboration – or love, if one wants to think of it that way – is mediated through the stomach.

3

The Honey Factory Production Line

A man sits on a chair in a small wood-panelled room and reads a newspaper. His nose and mouth are covered with a breathing mask. A tube leads from the mask to a small box on a plastic sheet that is also the lid of a beehive. This is how one of the more unusual products of the honey factory is applied. The product is air from the hive! For lengthy periods this man will breathe filtered air from the bee colony through a special inhaler. He is undergoing hive air therapy. This is believed to be beneficial for, among other ailments, lung and upper respiratory tract problems, because air from hives contains volatile antibacterial and antifungal substances. Even if the effectiveness of this microbiological treatment has not been confirmed, the combination of the pleasant olfactory experience and the soothing hum of the beehive alone would surely be therapeutic.

Hive air is a new product, but it shows that wax and honey are by no means the only substances produced from a honey factory that are of use to humans. Over the last decade, naturopaths have brought an entire palette of products from beehives into their practices. Apitherapy, with the catchphrase 'health from the hive', is now an established area of naturopathy. The medicinal applications of bee products are being explored internationally with some remarkable results, but to consider all the therapeutic uses proposed for propolis and pollen, bee venom and wax, royal jelly and honey is beyond the scope of this book.[1] Our interest here is on what bees produce and what they use it for, and we focus our attention on bees themselves, and the effect their labour and capabilities have on bee colony life. We can then look at what beekeepers do (or do not do) with all the honey factory products.

There is a fundamental dichotomy among honey factory products. Some come from the bees themselves, whereas others are processed from collected raw material. Bee venom, wax and royal jelly all come from the bee's own bodies. Propolis, bee bread and honey, on the other hand, come from raw material collected by bees and then processed by them. We look first at what bees can produce within their small bodies. We then turn to their resolute foraging activities.

What comes out of a honey bee

In summer, one is happy to finally kick off shoes and socks and feel the cool lawn underfoot – but then, a sharp, painful sting. With bad luck, this can result, fifteen minutes later, in a swollen foot. A bee was collecting pollen from clover, was alarmed at being trodden down into the lawn and so used her sting.

Bee venom: Painful, but good against rheumatism

All female bees, including the queen, possess a sting in their abdomens consisting of two glands, some muscles, a venom sac and the sting itself, which is composed of two setae, or spines. When a bee stings, first one spine penetrates the skin and then the other slides along the first, penetrating a little further. The two spines are alternately forced down until the entire length of the sting is embedded in the victim. Muscles then contract and inject the contents of the venom sac into the wound. About 0.1 milligrams of bee venom finds its way into the host. Apitoxin, the scientific name for bee venom, is a protein that induces an immune reaction in the host, accompanied by local inflammation, pain, swelling and itching. It is essentially harmless, if the host's immune system does not overreact. This is rare, but when it does occur it does so very soon after the sting. Large local swellings appear on the skin, the victim feels ill, the stomach is upset, the circulatory system collapses and breathing becomes difficult. The cause is the release of a flood of histamine from the host's body cells that can result in an anaphylactic shock. Untreated, this can be fatal.

Such a reaction is rare. More often, the sting leads to a swelling that increases over a few hours. Often diagnosed as an allergic reaction to bee venom, it is not infrequently treated with much fuss in an emergency ward. Is this really an allergic reaction? In a beekeeper's experience, stings are always painful but with time, contact with bees and many stings, the accompanying swelling becomes less or is absent. The immune system has adapted and no longer reacts so severely. Only stings in soft tissue, for example around the eyes – very painful – and lips, cause visible bumps. If bodies of beekeepers can become used to being stung, then it could be that the increasing occurrence of 'allergy' to bee venom is a phenomenon resulting from the growing distance between people and nature. Today, many people were never stung by bees in their childhood and their immune systems are untrained. In later years, an unfortunate confrontation with a bee on the lawn can lead to a painfully swollen foot. While uncomfortable for the victim, the consequences for the bee are far more severe. Sting setae are equipped with small barbs. These are not a problem if the bee stings another insect. The stung insect is at first held fast by the sting, while the bee injects its venom. However, when the venom sac is emptied, the bee can withdraw its sting and fly away. The venom sac is now empty and cannot be refilled, but the bee has done her job, perhaps as a guard defending the hive entrance against a thieving wasp. She would also attempt to withdraw her sting after using it on a vertebrate, but the sting barbs, embedded in the elastic tissue of a vertebrate, cannot be withdrawn. Instead, the bee tears the entire sting with muscles and glands out of her abdomen as she flies away, and dies.

Harvesting bee venom is complicated, because a bee can die after a single sting. Bee venom in most apiaries is therefore either in the venom sacs of bees or in the bodies of beekeepers. It is possible to collect venom by catching a bee at the hive entrance in tweezers, which makes her angry, holding her against a soft film and persuading her to sting that, a method that is not only irritating for the bee. A more usual way to collect apitoxin is to set up an electrically charged barrier across the hive entrance and give entering bees a small shock. They become aggressive and sting, squirting the venom onto a glass plate beneath the barrier, where it dries and can be scraped off and used in ointments for the treatment of rheumatic joints.

Royal jelly: Magic pudding for the queen

Bee venom is a secretion synthesised in bees' bodies from the third to the twentieth day of their lives. Once the venom sac is filled, the apitoxin synthesis ends and cannot be restarted after stinging. Production of royal jelly, another secretion we have already heard a lot about, is a different matter. This is made by the nurse bees from secretions of their 'milk' and mandibular (lower jaw) glands. Royal jelly is, as the name implies, the queen's food. She is fed royal jelly and nothing else as a larva and for the rest of her life. Naturopathy and the cosmetic industry make extravagant claims about the potency of this dull white, sour-tasting creamy substance for the prevention of facial wrinkles and in cancer therapy. It can be produced by motivating a colony to build and tend many queen cells, and we will hear later how this can be done. For now, though, once the queen cells are filled with royal jelly, they are removed from the hive and the contents scraped out of the cells. Freeze-dried, it can be used in medical and cosmetic products; most processed royal jelly stems from Asia or eastern Europe.

Beeswax: An aromatic building material for illumination

While bee venom and royal jelly are not products of the honey factory regularly harvested by beekeepers, with beeswax it is quite different. In the past, it was at least as valuable, if not more valuable, than honey. Beeswax candles were a high-tech luxury in times when there was no electricity and dark nights were brightened somewhat with oil lamps. Beeswax candles burned for a relatively long time, produced little soot and had a pleasant aroma. This last property was of some importance in a world where personal hygiene was not exactly prioritised.

Wax was nevertheless a rare commodity. As we heard earlier, wax was harvested with honey from bee colonies by breaking combs out of the hive, often killing the colony. Comb wax could be melted down and processed into candles after crushing honey from the cells. Honey extractors were invented later in the nineteenth century. A scarce raw material coupled with labour-intensive processing and manufacture results in an expensive product. Consequently, beeswax candles were found only among those who could afford them – in the residences of the aristocracy and in churches and cloisters. Bees are to be thanked for the special importance candles play in Christian religious services and liturgical feasts, Easter in particular. There, the bee's efforts are honoured. When blessing the Easter candle, the minister intones that it is 'prepared from the costly wax of bees'.

Today, reality and religious texts seldom agree and Easter is no exception. Easter candles, like most other candles today, are made from stearin. Beeswax rapidly lost its importance as a raw material following the discovery of stearin, although with the development of removable combs came the possibility of harvesting honey without destroying the combs. But empty combs are often given back to bees, and the amount of wax available in an apiary for other purposes decreased.

Beekeepers nowadays harvest mainly 'cap wax' and wax from old combs. Cell cap wax is a residue after harvesting honey. More about this below. Old combs are those in which young bees have been bred over two or three summers. A young bee that emerges from her cell leaves the thin wall of the pupal case behind her. These accumulate as brood cells are reused, the cell volume decreases and bees that emerge are smaller. The combs also become darker in colour. New brood combs are light yellow, but turn black after three summers, when it is high time to remove and melt them down.

Beekeepers can sell wax from comb cell caps and from old brood combs, and candles can be made from these. Beeswax also finds its uses in pharmaceutical products as an additive to ointments and as a protective finish for wood workers. Most beekeepers exchange old wax for new comb foundations, which are then mounted into frames and replaced in the hive for bees to build up into honey and brood combs. Many apiarists recycle their own wax and have the new comb foundations cast from their own old wax, or do it themselves. They can then be certain that there are no unwanted contaminants in the wax.

Collected resources: Propolis, pollen and bee bread

In Chapter 1 we explained how bees extrude wax from glands on their abdomens while building their combs. Wax is a building material that they produce themselves, in contrast to propolis, pollen and bee bread, which are collected resources.

Propolis: Duct tape from flower buds

Propolis is an important material employed in honey factories. Among other functions, it serves the same purpose as adhesive tape in mediocre hobby workshops: that is, to fix anything that leaks or is loose. The word 'propolis' is derived from the Greek *pro*, meaning 'before', and *polis*, meaning 'city'. Books on beekeeping explain that the name comes from the extensive use bees make of propolis to narrow the hive entrances: in other words, 'before the city'. Very early beekeepers had observed this behaviour: hence, the Greek name. It could certainly be true that in ancient times bees reduced the size of the entrances to their hives in pottery pipes with propolis. But today's bees living in magazine hives seldom do so; in most cases, they enlarge the entrances to prevent the hives from becoming too warm.

The word *pro* can also be more loosely translated as 'for', or 'for the protection of'. This comes much closer to the more general use of this unique material. It is a multipurpose compound that bees use 'for' and 'for the protection of' their 'city'.

Bees cannot produce propolis themselves. Instead, they collect resin from trees and plants and process this into propolis. The sticky substance that surrounds the buds of leaves and the flowers of chestnut trees, for example, is used to make propolis. Foragers take the resin and mix it with a secretion from their mandibular glands. Back in the hive, the already partly processed resin is combined with wax and kneaded again with their mouthparts. A pliable, intensely smelling red-brown mass forms in the warm hive, possessing a number of properties that are optimal for use in the colony. Propolis is an excellent adhesive, becoming hard and brittle when cold. Bees use it to fix and stabilise anything that is loose in their hives. Beekeepers could sing a song about this, and the hive tool is their favourite implement for good reason. It is as indispensable as a chisel and lever to pry free frame edges, which bees often bind firmly to the walls of the hive with propolis. Supers are also firmly cemented together with propolis, and the first checks of colonies in spring often pose a particular challenge to one's physical strength.

Propolis is water-resistant, although in a complex fashion of advantage to bees. Bees that make their homes in hollow trees plaster the inside of these entirely with propolis. Although resistant to water in its liquid form, recent research has shown that in hollow trees occupied by bees, propolis is an important factor in lowering humidity.[2] Just how propolis does this is unclear, and further investigation is needed. Some honey bees use propolis to build firewalls. Colonies of the Cape honey bee, *Apis mellifera capensis*, build their nests close to the ground among rocks and, apart from a few small entry holes, close off the entrance completely with propolis. Colonies survive naturally occurring bush fires behind these protective walls.[3]

In addition to being a multipurpose adhesive, wall paint and fireproof putty, propolis is also an effective disinfectant. It kills off bacteria, fungus spores and even viruses. With the help of propolis, bees ensure that the factory, despite its rather grubby appearance, is hygienically clean. Beekeepers sometimes find impressive examples of the effectiveness of propolis. Large insects and even mice can force their way into hives, are unable to find their way out, and die there. Because the corpses are too large for bees to carry out of the hive, and would rot if left, they are completely encased in propolis. Mummified bodies of moths, snails and mice are at times found on the honey factory floor.

This capacity of propolis was known to the ancient Egyptians, who used it, as in bee colonies, to embalm the dead. The disinfectant and healing properties of propolis were also known at that time. It is these properties that play an important role in its application today in apitherapy. Added to ointments, propolis supports wound healing; dissolved in the mouth, it soothes infections of the upper respiratory tract; swallowed, it is helpful against inflammation of the stomach and digestive tract.

Beekeepers collect propolis by scraping it off the frames and super. The raw propolis must then be dissolved in alcohol, filtered and rendered into a viscous mass by evaporating the alcohol. Freezing this produces a brittle material that can be ground to powder and added to ointments or other carriers. However, caution is advised. To make their propolis, bees collect it from wherever they find it. They have no interest in the production of a standard combination of different types of resin. The composition of propolis therefore reflects the area where the bees forage, their choices, and the time of the year. Medically, this can lead to problems, because bees sometimes collect resin from plants to which people are allergic. It is wise to use propolis preparations with care and from certified producers who usually obtain their supplies from apiaries specialising in propolis production. To collect propolis, a fine mesh grid is placed over the hive in place of a lid. Bees promptly fill the holes in the grid with propolis to make the hive weatherproof. The grid is then removed from the hive, frozen and the propolis pressed out and processed.

Pollen and bee bread

Harvesting pollen, another honey factory product, is tricky. Pollen consists of tiny grains that most flowering plants produce as male germ cells. It has already been mentioned many times in this book and, after honey, it is the most important source of energy for the honey factory. While honey offers bees energy as carbohydrate, pollen provides them with protein and fat, vitamins and trace elements, all of which bees need to survive and feed their brood.

Bees obtain pollen by dusting it off flowers. To reach the nectar in the flower cup, bees usually have to creep past the flower's pollen-carrying stamens, and powder themselves with it as they push by. Some plants, like balsam, have evolved mechanisms to press pollen-bearing stamens down onto bees. Visitors to this plant return to the hive with yellow 'rally stripes' stamped onto their backs. Pollen is not gathered only by bees collecting nectar. In spring, when the colony needs a lot of protein to support its growth, some foragers concentrate entirely on pollen collection. Bees can be seen literally bathing in willow blossoms. Flying home, they brush the sticky pollen from their bodies into bundles held in small baskets of bristles on their hind legs and arrive at the hive wearing small yellow trousers.

Beekeepers harvest pollen in spring when bees collect a great deal of it. A special grid is fastened across the entrance to the hive equipped with a collection container. Returning bees with pollen have to squeeze through the grid to get into the hive, and while doing so the pollen is scraped off their legs and falls into the container. The pollen has to be carefully dried and can then be used.

Pollen is considered by apitherapists and some athletes to be an unusually effective food additive. Rich in protein, trace elements and vitamins, it is believed to have positive effects on overall bodily condition, digestion, mental health and the immune system. But pollen can induce allergic reactions and is not easily digested. Pollen grains are covered in a tough protective shell that makes access to its contents difficult.

Hence, it is better to first break the pollen grains open, and bees achieve this with a special enzyme from their saliva, glandular secretions and a little honey. Pollen that is not immediately fed to larvae is mixed with honey, enzymes and secretions, stored in comb cells and known as bee bread, or perga. Perga is supposed to have undergone a fermentation process and so differ from fresh pollen although new studies suggest that this is not the case.[4] Nevertheless, bees prefer fresh pollen to perga if they are given the choice.

Bee bread can also be harvested by beekeepers if the colony has a surplus, although it is not easily removed from the cells. A simpler way is to carefully mix fresh pollen with honey and set this aside in a cool dark place for fourteen days. The result is a pollen paste that is close to bee bread and easier for humans to digest.

The product from which the honey factory gets its name is, unlike pollen, easily digested by humans and provides an immediate source of energy. We will now consider it: the food of gods.

The crème de la crème: Honey

Some people make an amazing fuss over wine. They decant it, taste it, discuss the year, location, grapes, tannin content, body and aroma, and eventually it expires. Interest is always piqued by the uniqueness of a particularly special vintage.

You might expect there to be a different attitude about honey, the most important product of honey factories. Buying rapeseed (canola) honey in a supermarket does not normally lead to any surprises. It is always clear and sweet. And forest, or honeydew, honey is always dark, slightly sour and runny. It might get more interesting when buying honey directly from an apiary in northern Germany, where bees have foraged in buckwheat. This honey is rare, often packed in small jars and has an acquired taste: it's as dark as beetroot sugar syrup, and has a strong aroma and a taste almost like liquorice. It's not the sort of honey you spread on bread: rather, it is a connoisseur honey for those seeking a special drop to enjoy in small quantities.

Raw materials and how bees process them

What is true for wine is also true for honey. Colonies living in the same location and foraging in the same area over many years produce honey that differs each year in consistency, colour and taste. Colonies living in the diverse vegetation found in and around residential area never produce the same honey. Honey from one hive can be light-coloured and sweet, with a mild aroma; that from another, 2 kilometres away, can be dark, slightly acid and spicy. How does this come about?

Two factors lead to differences between honey. The appearance, taste and consistency of the end product are determined first by how bees gather nectar, and second by the material they use.

The raw material from which bees make honey is the nectar they collect from flowers. Nectar is a watery, sugar-containing and perfumed substance that plants secrete from glands to attract pollinating insects. Insects need the energy-rich liquid for their nourishment and take it up from the plants. In return, they transfer pollen from flower to flower to ensure fertilisation and fruiting. Nectar is not the only raw material from which the honey is made. Bees also use an excretion from aphids, very different from the nectar of flowers.

Aphids are found on many plants and can cause flower fanciers and gardeners to despair. Bees love them because the small insects have a problem. They do not feed on nectar the plants produce in their flowers; instead, they penetrate the veins of leaves with their proboscises (long, sucking mouthparts) and gain direct access to sugar-containing plant sap. Aphids need sugar to survive but also require enough water, trace elements and protein. Plant sap does not contain very much of this and aphids have to consume a lot of sap to meet their requirements, and along with this acquire more sugar than they need. Most of this runs directly through them and is excreted as small droplets. In years when there is a massive infestation of aphids, a sticky layer of sugary aphid excretions – also, more delicately, called honeydew – covers the leaf surfaces of some trees. If there is enough of this deposit to warrant a visit, foraging bees gather honeydew by licking it up off leaves. Honey made from this is known as honeydew honey, or forest honey.

How bees collect nectar

When and how do bees find out that a source of nectar is worth visiting? To answer this question, we have to look a little closer at how bees collect nectar, the essential raw material for honey. You might imagine that, as depicted in children's television programs, a young forager leaves the hive with her small bucket and, humming happily, wanders from flower to flower, returning in the evening after many adventures with a bucket full of honey to the hive, where the queen accepts it with a kindly smile. Not quite true. We have already heard that the queen accepts nothing apart from her own special food, and foragers do not usually carry small buckets. They have a honey stomach to transport the nectar they consume. They also do not immediately leave the hive in the morning. Instead, they hang around a little, perhaps take a small nap, and wait for information. They do not go out looking for nectar as bumblebees do, hoping to find a source by chance. This is the job of scout bees. When scouts find a source, they mark it with their scent, collect a sample of nectar and fly home. There, field bees are given their instructions. They note the type of flower from the perfume carried on the scout's body and taste the nectar she has brought with her. And they attend her dance, which, somewhat simplified, proceeds as follows: the scout moves over the comb, describing a circle, although after completing only half of this, she turns sharply towards the centre and now moves slowly forwards, before turning sharply in the opposite direction to complete the other half of the original circle, bringing her back to the first sharp turn. The path she describes resembles a flattened figure eight on its side (see Figure 8). The 'straight' path through the centre of the two semicircles lies at a particular angle in relation to vertical. If the straight path of the dancer is directed vertically up the comb, it means 'fly towards the sun'. If directed to the left of vertical, it means 'fly left at this angle in relation to the sun'. If directed to the right of vertical, it means 'fly right at this angle in relation to the sun'. Dances are held on the surfaces of combs and we need to keep in mind that these hang vertically in the hive. During her passage through the centre of the figure, the dancer waggles her abdomen back and forth, while buzzing intermittently. She dances to a special rhythm – swing to the left, buzz, swing to the right, buzz, and so on. The frequency of the buzzing is about 250 hertz, optimal for signal conduction across the comb. The dance cannot be seen by her sisters, because it is completely dark in the hive, but they can detect the vibrations in the comb under their feet, approach the dancer and follow her on her way around, deriving an indication of the approximate direction they should take. This is only approximate and related to the sun's position seen from the hive entrance. The waggle run duration – that is, the short path through the centre of the figure eight – gives an estimation of the distance to the food source. A short waggle run means 'right in front of the door, girls'; a longer waggle run means 'you will have to fly for a while but it is worth it'; a slightly longer waggle run means 'quite a long way to go but still worthwhile'. The quality of the source is also signalled. If the scout bee moves slowly around the semicircle at the start of the waggle run, she signals 'try it, the quality is so-so'. A rapid semicircular run means 'amazing source'.

Dance followers that allow themselves to be persuaded by the information they have received set off to search for the food source. Should they find it, they collect nectar, return to the hive and dance. A trend then develops from information brought in by the first scout. More and more foragers in the hive receive information about the kind of nectar available and its location from more and more foragers that have already visited the source. Gradually, an air traffic corridor forms between hive and food source. Bees that have not experienced the dances at all now also join the flow of traffic.[5]

On a good day, with fine weather, many scouts are underway and may simultaneously discover different sources. Dance floors in the hive become parties held in different areas of the comb, advertising locations where there is something to be had. Recruits follow the dancers, meet them at the food source or even on the way there, return to the hive and report their own successes. Others are persuaded and join in. Gradually, field bees gravitate to the source that attracts most recruits.

The most attractive nectar sources are so near or so rich that foragers bring in much more nectar than they use to fly there and back. The important consequence stemming from this convergence of information with regard to profitable food sources is that forager bee efforts are highly economically employed, resulting in impressive optimisation of cost and effect. At times, on an early summer morning and even later, there can be very little activity at the hive entrance despite the best possible weather. Bees stay home and do not waste their energy searching. And then, only an hour or so later, one can hear buzzing from some distance away and the hive entrance is crowded. The temperature has perhaps increased by one or two degrees, or maybe linden trees have finally managed to fill their flowers with nectar. Nothing now holds bees back and a strong colony can easily bring several kilograms of nectar into the hive. Bees optimise the cost and effect of a hive by mobilising their workforce precisely when it is most profitable. A relevant component of this efficiency is the concentration of foragers at the best food sources, brought about by the convergence of information in the hive and resulting in flower constancy. Suppose trees in an apple orchard 300 metres southwest of the hive begin to bloom on a Monday and bees have flown back and forth to collect from these trees. On Tuesday, the bees that had visited on Monday will visit again to see if there is still something left to gather. They do not need more information; they already know where the orchard lies. They will still know where to fly even if bad weather prevents their leaving the hive for an entire week. Foragers only need new information when there are no more flowers on the trees and no nectar to collect.

How bees find their way to the dancers

The recruitment of foragers is an intricate behavioural chain that begins in the dark hive. It is here that scouts and foragers who have been to food sites must attract the attention of dance followers and pass on to them information about the site. But how can one gain attention in the dark? The 'house telephone', mentioned in Chapter 1, plays a critical role here.

Every movement an animal makes is associated with measureable changes in physical forces and fields in their vicinity. Airstreams and oscillations caused mainly by wing vibration accompany dancing honey bees during certain phases of the waggle run. The waggle motion and wing movements are transferred as vibrations to the comb. Furthermore, dancers usually have a higher temperature than other bees around them, and finally, the dancers exude a particular odour.

The fact that such physical–chemical events can be measured does not automatically mean they have any importance for the behaviour of animals. One can only determine if and how they are relevant by watching the behaviour itself.

Modern technology has been crucial in obtaining better insights. Slow-motion videos taken within the dark hive are an important tool, allowing an analysis of the 'microbehavioural' details of waggle dancers and followers. Recordings of dance scenes can also be rewound, making it possible to determine from which direction, and from how far away, a follower is attracted and how she behaves when she first becomes aware of the dancer. Behavioural sequences recorded on videos show that a bee first turns her head in the direction of the dancer before she moves. She has detected, through vibrations, odour or other signals, that one of her sisters has some interesting information to share. After turning her head, the follower bee moves towards the dancer and, when close enough, she touches her with her antennae. The follower, now in contact with the dancer, follows her around the dance path.

Recruits have occasionally been seen to follow dancers when standing on a comb directly across the passage between combs and so standing back-to-back with the dancer. These followers are not attracted from a distance like those standing on the same comb as the dancer. They also do not turn their heads towards the dancer, nor do they move actively towards her. Back-to-back followers join a dancer only if they happen to touch her with their antennae and so detect the activity. The different behaviour of the followers can be explained if we take into account that back-to-back followers are not standing on the same comb as the dancer. She receives no comb vibrations because these are not transmitted across the gap between the two combs.

Honey varieties

Focused foraging results in bee colonies producing very different kinds of honey from year to year due to the mix of nectar from the best sources in the area. Perhaps in one year bees exploit a rich honey flow from dandelions. The honey would have been yellow and aromatic. In the following year it rained during blossoming of dandelions but chestnuts happened to be particularly good. Harvested honey would have been darker and with a slightly spicy taste, if mixed with nectar from earlier flowering fruit trees.

Beekeepers make use of flower constancy when they travel with their hives to harvest different kinds of honey. Travelling beekeepers do not leave their hives at a single location and let the bees collect whatever is there for the total flowering period; instead, they take the hives to places where good nectar sources are expected. A colony's journey can begin in April, when they are set down in a field of canola while it blooms and bees have plenty of nectar to collect. The beekeeper harvests canola honey. When this is over, the colony can be moved to where locust trees are now in full flower. Next, to linden trees from the middle of June to the middle of July, and in August either to blossoming heather or into forests where perhaps the trees have something to offer – with luck, perhaps a good supply of honeydew.

THE HONEY FACTORY PRODUCTION LINE

Such journeys are challenging for both bees and beekeepers. The transport and care of colonies between widely separated regions requires a full sixteen-hour day. And frequent changes of location are not good for bees. Colonies are subjected to enormous stress because hives are sometimes in transport for long periods. Imagine if one's home suffered mild earthquakes for hours on end. Nothing is destroyed, but the continuous shaking would surely be hard to bear. And then there is a period when bees have to be 'flown in' at a new site to get to know the structure of the area around their hive. Travelling bees are relocated overnight when they are all at home, and next morning when they come to the entrance they are faced with a totally strange environment. This is discouraging and results in considerable effort being needed, because it takes foragers a couple of days to accustom themselves to a new area. But the biggest problem for travelling apiaries is a lack of diversity in the bees' diet. Areas that offer large honey flows, such as canola or heather, provide only one kind of pollen, which, apart from supplying the essential protein and fat, is an important source of vitamins and trace elements. Bees need a diverse diet composed of a mixture of pollen from different kinds of flowers to remain healthy.

Not many apiarists impose the sort of honey flow pursuit described above on their colonies. Most travel only during canola blooming, but otherwise bees have a fixed home site where they can access a healthy mix of many different kinds of flowers. Alternatively, some colonies are set down near canola, others near linden trees and still others in forests and then left there. That way, beekeepers can still harvest different kinds of honey without stressing their colonies.

How nectar becomes honey with patience and saliva

How does nectar, this watery sugar solution, become honey? Simple enough, one may think. After all, nectar is sweet and aromatic, just too diluted. So bees reduce the water content and there it is. Is honey not just thickened nectar?

This is not entirely wrong, and one of the primary tasks of honey-maker bees is to reduce the water content of nectar. Foragers have already started the process. Flying back to the hive with their honey stomachs full of nectar, they regurgitate small drops of nectar and hold this in their jaws. The passing air evaporates some water from the nectar they carry before it reaches the hive, where the process is repeated. Honey makers take nectar up out of cells, regurgitate it and spread it thinly in new cells to increase the surface area and promote evaporation. The nectar becomes thicker and is transported from the lower to upper regions of the hive, because bees prefer to store their honey above the brood nest. Completed, or 'ripe' honey as apiarists call it, ideally has a water content of 18 per cent. Once it has reached this level, bees cap the cells in which it is stored with an airtight wax lid. A comb filled with ripe honey has all the cells capped.

Ripe honey is much more than just thickened nectar, and an important component of honey making, in addition to the bees' patient labour, is their saliva. Foragers mix their enzyme-containing saliva with nectar and start the biochemical conversion of the sugars contained in nectar. Sucrose is split into dextrose and fructose. Bees contribute acids, protein and other enzymes to honey and add inhibin to inhibit the growth of yeasts and fungi. In comparison to nectar, ripe honey contains less water and has a considerably longer shelf life. Its low water content, high sugar concentration and the presence of inhibin prevents honey from fermenting or becoming covered with mould. Stored in a dry, dark and cool place, it keeps practically forever. Beekeepers tell stories about honey placed in the sarcophagus of an Egyptian pharaoh, which is supposed to have been enjoyed by the archaeologists who discovered it.

Humidity in the hive: Bees and steam

Bees evaporate water from nectar when they turn it into honey, rather like we do when we hang washing out to dry. The surrounding air absorbs water, provided it is not already saturated with it. How do bees control humidity in their hives?

Humidity is the proportion of water vapour in air and is primarily dependent on temperature. The warmer the air, the more water vapour it can contain. A cubic metre of air at 16°C holds about 6 grams of water, and at 35°C, the temperature of brood combs, a cubic metre holds 40 grams of water.

Nectar, with a water content of about 60 per cent of its weight, is a source of moisture in hives. The metabolism of bees and the brood also produces water. Bees heat honey cells when making honey and reduce its water content from 20 per cent to 16 per cent. The warm air produced by heating the cells can absorb water released from honey, but this also raises the humidity level in the hive and creates some real problems for bees.

Honey processing comes to a stop if air in the hive is saturated and can no longer absorb water from nectar. Worse, honey is hygroscopic and actively absorbs water. Already processed honey in a damp hive starts to increase its water content, leading to the danger of fungus, yeast contagion and fermentation.

The only remedy is to replace damp air in the hive with drier air from outside. In summer, a frequent sight is the presence of many bees sitting with their heads pointing to the hive entrance and fluttering their wings with abdomens raised. The resulting airstream directs cooler air into the hive. Is it the increase in humidity in the hive that releases this behaviour? Or do the bees always heat and ventilate when nectar is being processed?

A fully occupied honey beehive is a relatively large and complex system in which to experimentally regulate and measure humidity in all passages and corners. The simpler nests of bumblebees offer a practical alternative in which to explore humidity control and then extrapolate the results to honey bees.

Bumblebee nests can be accommodated in half a shoebox. When the nest is provided with only a single narrow and short entrance tunnel, a bumblebee is able to replace the air in the nest by standing in the tunnel and fanning its wings. Artificially raising the temperature of the nest triggers this behaviour. Replacing air in the nest with a damp atmosphere, but without changing the temperature, surprisingly, does not induce bumblebees to ventilate.[6] Honey bees and bumblebees both have special receptors on their antennae that are sensitive to humidity and are capable of determining its extent. But active regulation of this parameter is apparently not included in their natural behavioural make-up. The evolutionary development of bees may offer an explanation here.

Nature is economical. When bees do not react to a massive increase in humidity in their nests, it could mean that they are not aware of it. However, we can exclude this option because we know that they have receptors that are sensitive to humidity. Alternatively, it could be speculated that under the natural conditions in which bees have evolved such events are very rare and development of an appropriate behaviour would be without evolutionary advantage. Measurements taken from within tree hollows occupied by bees and their combs provide some interesting clues. Water condensation from saturated air is never found in such hives because the relative humidity never reaches the point at which water condenses into droplets. Instead, relative humidity in these hives tends to be the opposite of that outside: high during the day when it is low outside, and low inside at night. There is no need to evolve a behaviour to avoid the formation of condensed water in a hive when it never occurs in natural nests. Bees fanning their wings in front of the nest are cooling the hive, not drying it out.

Industrious bees and the beekeeper's contribution

Apiarists know the time to harvest honey has arrived after bees have dried the honey and closed the honeycomb cells. The removable frames and supers in modern honey factories allow harvesting to proceed without mistreating the bees.

Bees ripen and store their honey above the brood nest, and beekeepers exploit this tendency, separating the hive into a brood area and a honey area. These two regions, depending on the frames, the size of the colony and the honey flow, each consist of one or two cases and are separated by a grid, the *queen excluder*. This is a stable wire grid laid on top of the brood nest case to separate it from the honey supers above. The grid is constructed so that spaces between the wires allow workers to slip through but are too small to let through solidly built drones, or the queen with her abdomen swollen with eggs. Honey makers transport ripe honey up into the area above the brood when the nectar supply is strong and there are so many active field bees that more nectar is collected than is needed to feed the brood. The grid prevents workers from escorting the queen up into the honey area to lay eggs, and so only honey and nothing else is now stored there. A strong colony can fill a honey super in a few days, given a good nectar supply and good weather. The honey ripens about four to six weeks after the beginning of the honey flow. If dandelions and fruit trees begin to bloom in the middle of April, beekeepers can harvest spring honey at the end of May or beginning of June.

To further avoid stressing bees when harvesting honey, beekeepers have devised a procedure that reduces their number in the supers. Before harvesting, they lift the super, remove and replace the queen excluder with a board with a few holes in it just big enough for workers to slip through one at a time, and replace the super. The board limits circulation of the air in the super and the concentration of the queen pheromone, which holds the colony together, sinks lower and lower in the honey area. Honey makers there are disturbed. Has the queen died? Is the colony in danger? They set off in search of Her Majesty, following the little scent that drifts through the holes in the board. One after the other they leave the honey area and move into the brood nest. Twenty-four hours later the super is virtually empty of bees and the beekeeper can remove the super, withdraw the combs, and brush any remaining bees off the combs near the hive entrance. From here, they find their way back into the hive. The bee-less super is now taken home to the separator.

The honey separator

The first step in separating is to employ a special comb or a long-bladed knife to remove the wax caps from the cells. This is the 'cap wax' that we mentioned earlier. Opened honeycombs are placed into the separator, which is basically a drum containing four vertical sieve plates mounted on a structure that rotates within the drum. Frames with the honeycombs are placed vertically in the rotating structure with one side against the sieve. The combs and sieves are then rotated with a crank or motor around the central axis of the drum. Honey flies out of comb cells under centrifugal force, through the sieves and onto the outer walls of the drum and drains down through a funnel into a collecting bucket. Combs have two sides but cannot be emptied sequentially because if cells facing the sieves are completely clean, the weight of honey on the inside face of the comb would break through the comb and destroy it. Instead, the cells have to be gradually emptied by taking the frames out, turning them around and alternately spinning first one side and then the other. Three changes are usually enough to produce combs that are empty, still intact, and can be returned to the colony. The beekeeper removes the board over the brood area, reinstalls the queen excluder and sets the super with empty combs in their frames back onto the hive. In good times, bees will have refilled the super in four to six weeks.

Processing the honey

The honey that bees have so carefully prepared is now in the hands of the beekeeper. It is now the beekeeper's turn to work, for honey that comes out of the comb cannot be directly poured into jars. Small impurities get into honey during separating. This is not serious, as no-one was ever harmed by a few particles of wax, but it does not look good to have wax particles floating around in honey, and it also sticks to teeth. Honey has to be sieved, and this takes place during separating, before it reaches the collection bucket, and most wax particles and a few bee legs are filtered out.

The honey foams in the collection bucket and this spoils its appearance, although it is not a strong bubbling foam that indicates fermentation! Perhaps it is just small air bubbles that entered during separation. Also, proteins form a whitish film over the honey, and this is carefully skimmed off so that the surface is again clear.

Most honey becomes cloudy not long after being separated. This is not a sign that it is going bad. Ripe honey begins to 'candy' after being harvested, when small sugar crystals form. This is a completely natural process that depends on the proportion of the different sugar types within the honey. Honey from flowers usually has a high proportion of dextran and candies soon after separating. Honey from honeydew contains more fructose and candies later, and sometimes not at all, or small crystals are found floating in the clear honey and at the bottom of the jar.

Crystallisation is a nuisance only when crystals become too large. Linden honey can get so hard that one almost needs a hammer and chisel to get it out of the jar. There is a simple way to prevent this – stirring! The moment crystals appear, the honey is stirred slowly for five minutes each day for a few consecutive days with an electric hand drill and a special accessory for honey stirring. Large crystals rub against one another during stirring and gradually grind each other down, just as shaking a tin can filled with sugar cubes for a while will result in fine sugar. When the surface of the honey begins to appear opalescent after a few days of stirring, this is the signal to stop. There are so many fine crystals now present that the honey will no longer harden, and it stays easy to spread. Now it can go into a jar and from there onto a slice of bread.

Winter feeding

Beekeepers who take the last honey harvest of the year from their bees create a problem for the bees and for themselves: the bees have no reserves for winter. There are still a few flowers with blossoms in late summer and honeydew is plentiful if the aphid population is high. But an excess with which to restock the honey store is no longer available. There is barely enough for day-to-day living.

The bees of apiarists who do not provide an alternative to the honey they extract will starve at the latest in spring, when food is needed for new brood. For this reason, colonies are hand-fed from the end of July to the end of September. The queen excluder is removed so she can move freely through the hive, and the honey area is accessible to all. There is now space above the brood nest, the queen lays fewer eggs and after about three weeks the lower floor of hive is without brood. There remain only empty cells, on which disinterested field bees hang around. The beekeepers remove this case, add a frameless *feeder super* above the original honey area, and prepare a small paradise for the bees. Once the feeder super is in place, they supply it with liquid food – a mixture of fructose and dextrose with a water content of about 20 per cent. This is a viscous and very sweet sugar solution. Access to the feeder is through a small slit in the base and once through this and over a small wall, bees find a sweet sea directly over their heads. A grid prevents them from falling in and drowning. Soon they can be seen standing close together and licking up the sugary syrup. A strong colony can transfer and pack 15 kilograms of food into comb cells and cap them with wax lids. The honey stores have been replenished and the colony's survival guaranteed.

Sugar syrup is not honey. Even if 98 per cent of honey is a mixture of different sugars, the remaining 2 per cent consists of over 180 other components, such as enzymes, minerals, trace elements and so on. Winter food is a rather thin soup in comparison to honey. There are studies suggesting that bees overwintering with honey instead of sugar are somewhat more vigorous in spring. On the other hand, honey as winter food can also pose problems for bees. In contrast to sugar solutions, honey contains ballast that collects in their gut. A colony that survives the winter on a ballast-rich diet such as honey from honeydew, and are prevented from leaving their hive by a cold spring, cannot fly out of the hive to purge their digestive tracts. They are forced into the unsanitary situation of emptying their gut onto the combs. Bacterial infection follows and the colony often dies. While sugar syrup is a poor substitute and not a real replacement for honey, bees do manage with it. And in fact, very few apiarists take all the honey they find in a colony. Combs at the edge of brood cells and small areas filled with honey from autumn honey flows, now more common in our latitudes due to climate change, are left for the bees. Main meals in the bee canteen in winter may be sugar syrup but now and then there is a honey dessert.

4

Founding a Daughter Company

The Swarm

It is the end of May and the day is sunny and warm. The hive entry is busy with foragers continually landing and taking off. Nothing exceptional, just normal activity on a day with a good honey flow. Towards midday, the situation suddenly changes and the hum from the hive swells in volume. The landing board at the hive entrance and the walls of the hive become hectic with activity. More and more bees flood out and take off, but instead of flying away, they form a large cloud in front of the hive. The friendly humming becomes a loud and furious buzzing, with bees circling in the air all around the hive.

And then she arrives. The queen emerges onto the landing board, pauses for a moment in the unfamiliar bright light, and takes off.

A few minutes later the situation changes again. The buzzing cloud of bees disperses, the noise diminishes and there are fewer bees in the air. The usual busy coming and going dominates the hive entry, but there are not as many bees departing and arriving as before. Several metres away from the hive, a fascinating spectacle meets the eye: thousands of bees hang together in a thick cluster suspended from a tree or in a hedge or bush. A swarm has left the hive to found a daughter colony. How did they come to this decision and what happens now?

Driven by instinct

The queen of a colony lays eggs, the nurse bees care for them and they hatch into larvae, pupate and finally become young bees. But somewhat surprisingly, this procedure does not produce more colonies! If only a single colony existed and the queen was immortal, laying eggs and nurturing the offspring would never lead to a second colony. But, of course, the queen is not immortal, so eventually a colony would disappear with her death – if there wasn't another way to form a new one.

Social insects have existed for several million years, and with the help of some evolutionary strategies they have propagated their colonies. Bees produce offspring to maintain their individual colonies, but they increase the number of colonies through *division*. A proportion of bees in the hive swarm out and start a second colony. A *swarm drive* exists in all colonies and, like many natural instincts, is often suppressed but when aroused is uncontrollable. The swarm drive in bees occurs in the springtime, usually within a two- to three-week period depending on the locale.

In spring and after the winter break, the colony begins to grow. Thousands of young bees emerge over a few weeks, with the arrival of the new brood and the spring blossoms. A colony that consisted of 5000 individuals in March could easily have 35,000 bees by the middle of May. Honey factories now run in top gear, tending the brood, collecting nectar and pollen, building combs and storing supplies. Everything happens at once and eventually the inevitable occurs: all the combs are filled with brood, pollen or honey and there is no space for more combs.

This is bad for morale. The queen finds nowhere to lay and has her rations cut by the bees feeding and caring for her. She does not stop laying completely, but produces far fewer than the usual 1200 eggs per day. The ovaries in her abdomen begin to atrophy and she slims down.

Fewer eggs and larvae mean less work for nest bees, which is also bad for the feeling in the hive. There are many nurse bees, food for larvae is available, but it is not needed. And the building gangs are no better off. They would like to build, but there is no space left in which to do so. The colony's potential is thwarted. The possibility of growth is there but it cannot be expressed. The bees soon notice it is time to start something new: the swarm drive is aroused and the colony prepares for division.

And what do they need? Correct, a new queen! A colony preparing to swarm builds a number of 'swarm cells': small round wax bowls that are constructed along the edge of a brood comb. Strong colonies will build ten or more of these, and the queen is led to them and lays an egg in each cell. Nurse bees recognise the form of these swarm cells, or queen cells, and feed the larvae that hatch from the eggs exclusively with royal jelly.

The swarm cells are capped about eight days after the eggs were laid. The small bowls have since become elongated and the larvae in them will pupate and metamorphose into queen bees.

Now is the time to swarm, and the colony does not wait for a young queen to emerge from her cell and then make off with her. Instead, it is the old queen who leaves the honey factory to found a new one. On a warm spring day, slim from a curtailed diet and able to fly again, she takes off with about half the company. Up to 20,000 bees now search for a new home for their queen.

The first leg of this journey is short. Although the queen can fly, she is not able to match her performance as a young queen on her nuptial flight and has to rest only a few metres from the hive. She settles, not far off the ground, on a branch or in a tree.

Her small court settles around her and immediately starts to release pheromones. The bees raise their abdomens, open their scent glands and spread the pheromones in the air around them by fanning their wings. The message to the swarming bees is 'over here, girls, the queen is waiting'. Gradually the bees receive the signal and form the swarm cluster we described earlier.

The situation of the company members that stayed behind in the hive is somewhat complicated. On the one hand, the bees who joined the new factory team are hanging in the tree; on the other, the staff who remained in the old works are waiting for the new queen to emerge and for what will happen after that. But first we will look at how the new settlers fare.

Swarm intelligence: How bees move house

A swarm is an impressive sight. Tens of thousands of bees leave the hive with the old queen and camp together in a tight cluster, often in a tree close to the old hive. A little while later, they quite suddenly take off, all together, and head in the direction of a new home site that may be several kilometres away.

Unlike birds hunting for nest sites, an entire swarm cannot fly about to inspect likely places before starting to build. How, then, does a swarm with thousands of individuals find a possible new home and then decide to accept it like a large single animal? And how do bees persuade 10,000 or more individuals to all move together to the new nest site? The answers to these questions draw attention once more to the fascinating communicative interaction within a bee colony that we still do not entirely understand.[1]

Scouts are sent out

Bees in a swarm first employ a strategy that early nomadic humans who were hunters and gatherers used when seeking a new home base. They sent scouts to survey the surrounding area. The information brought back to the group was then discussed and used as the basis of a decision on which way to go.

Honey bees use the same technique. A relatively small group of scouts fly out from the swarm to search for a new and suitable location. This has to happen promptly because the colony is vulnerable. A sudden thunderstorm could destroy them, and food reserves are minimal. Before swarming, bees take up as much food as they can hold in their honey stomachs, but this is not only for their own nourishment: it is the energy supply for comb wax in the new home and to feed the first brood. The longer a swarm cluster hangs around outside, the more precarious its future.

Scouts hurry off, and each one that finds what they feel is a likely spot becomes a real estate agent. She returns and dances on the surface of the swarm cluster, advertising what she has found. This dance, called the waggle dance, provides information about the approximate location of a possible home site in exactly the same way that foragers 'tell' the hive about nectar sources. The vigour and duration of the waggle dance reflects the enthusiasm of the scout. The more qualities the new hollow exhibits that have proved optimal for bees during their evolution, the more enthusiastically the bee dances. The size of the hollow is a deciding factor. It must be large enough to allow growth over years and have enough space for comb building, but also not be too large. Scout bees walk around in prospective hollows, following different paths across inside surfaces to gain an impression of their shape and volume. The new home must be dry; be well above ground level; and have an entrance that is not too large and, if possible, facing away from the prevailing wind.

Real estate agents persuade

The scouts' discoveries vary. Some find only poor shacks, others high-class villas. Now they have to inform the swarm cluster, convince them and move them to a decision. But how? A scout group of a few dozen bees dancing on the surface of the cluster can reach only a tiny fraction of bees in the swarm, although all of them must reach the goal together.

The task is achieved through a sequence of different behaviours within the swarm. First, it must be established which of the discovered alternatives is the best and the way to the new home must then be signposted. Finally, the few individuals that know the way have to persuade every bee in the cluster to break up and fly off together when the decision has been made.

The enthusiasm of scout bees is important for deciding which site will be chosen. Scouts that are not completely convinced by their own discoveries stop dancing when they detect more energetically dancing colleagues. In accordance with this simple principle, advertisement for less attractive sites falls away and eventually the only bees left dancing are those most convinced by what they have discovered. A decision about the new home has been reached.

Winners of the election now indicate the way to the new site to other individual bees. They dance on the cluster, and fly back and forth to the hollow, where they perform buzzing flights to mark it with pheromones. They may also lay an odour trail from the cluster to the new address. Scientists think that under the most natural conditions, swarming and searching for a new home would take place in forests, where such an odour trail would not be dispersed rapidly, but this is speculative: there is no solid supporting evidence for this theory. In any case, the scouts get others to join them in their behaviour. Recruits visit the site, return to the cluster, dance, fly to the site and mark it with scent.

The hive feels the beat

This process could continue until eventually all bees in the swarm are recruited and know where to go, but to individually recruit every member in the cluster would take far too long. The colony is defenceless and needs a new roof over their heads, and soon. How can one bring an entire swarm, surrounding and protecting their queen in their midst, to break up and fly en masse to a small opening in a single tree in a forest several kilometres away?

A behavioural change occurs once the number of bees that know the location of the goal reaches a certain size and have danced and marked it. Instead of dancing and flying back and forth, they now burrow down among the tightly packed bees in the swarm. A small microphone buried in the swarm records short piping tones that scout bees make with their flight muscles.[2] Bees are deaf and do not hear this piping but can feel it. The strong sound pulses activate sensilla on their antennae and heads and vibrations are also detected by their legs. One can imagine that bees detect these signals as we would those from a bass drum or guitar. We hear them but also feel them in our bodies. Within the tightly packed swarm, bees in contact with a scout would detect her piping with their entire bodies. Two microphones placed in a hive allows the path taken by the scout through the cluster to be reconstructed. Each piping tone, made as she forces her way through the crowd, lasts about half a second. Her path through the cluster is tortuous and can be up to half-a-metre long. Once at the surface again, she dances and re-enters the cluster for another round of piping. In between such bouts, she may fly to the new site and back.

Contact between the piping scouts and the swarm bees has a remarkable effect on them. The body temperature of bees within the cluster begins to rise once the piping begins[3] and following a scout with a heat-sensitive camera reveals a trail. The body temperature of the scout is higher than her sisters in the cluster who, evidently animated by her piping and the warmth of her body, begin to raise their own body temperatures. After about half an hour of energetic piping, the temperature of the swarm cluster reaches around 35°C.

And then it happens – the swarm cluster explodes. In a single stroke, the entire swarm takes off and forms a cloud in the air, which at first appears undecided about where to go. The scouts again play a decisive role, by flying rapidly back and forth through the swarm along a line oriented towards the goal. Slowly swarm bees move into formation, the cloud takes on the shape of a cigar and begins to move quite rapidly in the direction of the new home indicated by the scouts and their scent trails.

Soon, they arrive at the new site.

Figure 16

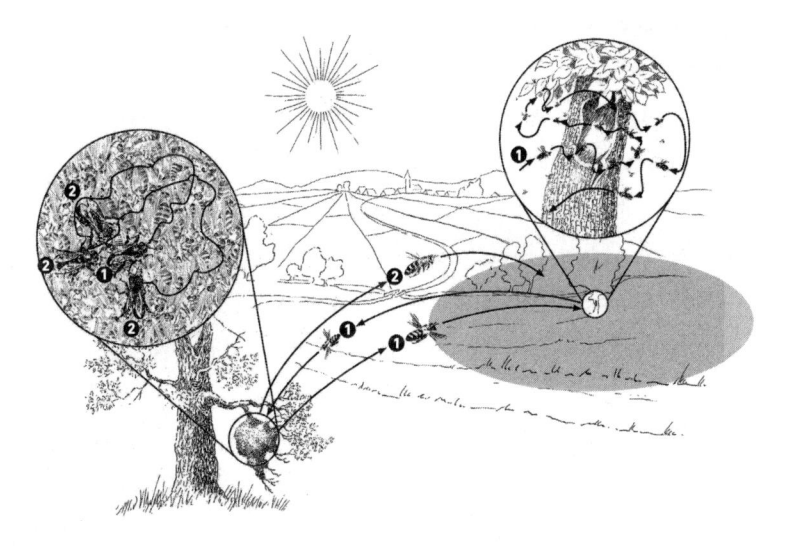

Recruitment to a new nest site. Bees employ the same communication and supporting signals they use to recruit bees to feeders (mini-swarms) (see Figure 13), to bring thousands of swarm bees precisely to a small location in the landscape (from Tautz 2015).

A superorganism with multiple talents

The new premises are at first completely empty. The staff are there, but the most important factory equipment, the combs, is lacking, and the provisions will not last forever.

To survive this critical period, the colony has to get busy and ensure that enough young bees emerge in time to lay up supplies for the coming winter. The impressive performance of the colony as a superorganism expresses itself in its capacity for 'plasticity'. Swarming changes the usual rules that determine the tasks that bees, especially the foragers, undertake at particular ages. Although almost at the end of their short summer lives, the foragers now join the building gangs and extrude wax from their abdomens for comb building. They are also active in caring for the brood; the brood combs in which the queen can lay are quickly constructed. In less than four weeks, the first young bees that never knew the old hive emerge to work in the new production halls.

The old queen has brought a new colony into being. From one colony, two have been created.

How beekeepers mediate new home sites

For bees in maintained honey factories, the process is simplified, and founding a new factory occurs quickly and smoothly.

As soon as a swarm has left the hive and gathered into a cluster in a tree, a beekeeper arrives to collect it. The procedure is straightforward. First, the swarm is lightly sprayed with water, which cools it down and dampens the bees' wings, making it harder for them to fly. The beekeeper then brushes the entire swarm into a box or basket which has a base with an entry. A lid is placed on top, and the box or basket is set down. With the queen in the box, any bees still flying around will soon find their way to her and enter the perfect home. That evening, once bees have stopped flying, the box can be sealed and kept cool overnight in a cellar. The next day, the bees are tipped out of the box into an empty hive with a baseboard, lid and entrance. Frames with centre partitions and comb foundations are slipped in among the milling crowd of bees, the lid is fitted and the new company moved to its new location. This cannot be too close to the old company, because field bees would then simply fly back to the old factory door they know so well. The new swarm would lose its vitality and be too weak to make it through the coming winter.

Events in the old factory: The terror of succession

In the meantime, what is going on in the old factory the swarm came from? Those who stayed behind do not appear to be disadvantaged. They have well-filled stores, and in the next few weeks many young bees will emerge from the cells where the old queen laid eggs. And the combs are already there.

The battle for the throne

But all is not calm. The old colony has no queen, and unrest and violence surface when there is no clear power structure in the hive.

Many queen cells were built during the swarm drive period, and sixteen days later these all contain princesses with one ambition: to be the sole heir to the throne.

So how do they cope with the competition? For a start, they talk.

Young queens about to emerge, press themselves against the walls of their cells, produce a 'quacking' noise with their thoracic musculature and wait for a reply. If the colony is still in swarm mode and another young queen has already emerged, she answers with a 'tooting' noise. The answer to the quacking queen still in her cell is: 'Stay where you are, I have to swarm and will be gone soon.' Fairly often, there will be a second swarm, or *afterswarm*, where a young, unfertilised queen makes off with half of those still left after the first exodus.

The first goal of the remaining young queens that emerge after the swarm drive has ebbed is to eliminate the competition. This is simple enough if they are still in their cells. The emerged queen gnaws a hole in the wall of a queen cell, inserts her abdomen and stings the occupant, who dies. A duel to the death on the surface of the comb is the only way to deal with those already emerged.

At the end of this period of unrest, the surviving queen comes into season and soon takes off on her nuptial flight. She returns well-served to the hive and begins to lay, and although the old colony has significantly fewer bees than before, it now has a hive mother. Equipped as it is with a fully functioning factory and enough food stores, the colony will certainly survive the coming winter.

Hive division is ingenious from an evolutionary perspective. The old, alreadly fertilised queen who leaves the factory can immediately begin laying eggs and quickly establish a new colony. Members of the colony who are left behind must wait up to five weeks for a new queen to produce offspring, but can live through this period because they have enough food and the absence of the foraging field bees who left with the swarm is of no great consequence. The new young queen begins her reign under the best possible conditions. In principle, the old bee colony, provided with a new queen who can lay prolifically, is immortal.

Afterswarms: Poor and homeless

Swarms that leave the hive after the first main exodus are in a predicament. Apart from the honey in their stomachs, they have nothing: no home and an unfertilised queen. Their only advantage is that the young queen has no heavily developed ovaries and can fly as well as the workers. Such swarms, known as afterswarms, are recognisable because they are often very small, form clusters high in trees and usually soon fly off, making them difficult to collect. Perhaps these swarms manage to open a small workshop far away from the original hive. Perhaps, like pioneers in the wilderness, they can settle where there are no other bees – but there is no certainty. Perhaps in times when wild forests and savannahs remained features of the landscape, it was an advantage for colonies to be able to fly far and fast. Bees could then repopulate areas such as burned out forests. Today it is difficult for afterswarms to find an appropriate home site. And even if they manage this, the queen still has to complete her nuptial flight and return safely to the hive.

Killjoy beekeepers

Wherever there are compulsive drives, there are also moralists who want to control and suppress them. Beekeepers take on this role. As we saw in Chapter 1, the early basket apiarists encouraged swarming because the more colonies that swarmed, the more hives they had to set out in the heather in late summer. But with the invention of magazine hives, swarming was no longer desired. Now the goal was for colonies to become large early in the year, so that the entire summer's supply of nectar could be collected and all the different honey flows processed into honey. A beekeeper's task is, as far as possible, to prevent hives from splitting and forming daughter companies. How is this done?

Those with work to do make no mischief

The swarm drive is aroused when potential in the hive is blocked, and this happens quickly when the honey factories run out of space. With enough space to live and work in, colonies seldom swarm, so one method to control the drive to swarm is to provide the hive with enough space. In Chapter 1 we described how simply adding another super to the hive can take care of this. Here, bees can find new frames complete with comb cell foundations. Bees can quickly build up cell walls, the queen can lay and nurse bees soon have many larvae to tend. Two supers and a queen excluder is usually enough space for an entire season.

However, colonies will still swarm despite the adequate space; sometimes a change in the weather is enough to make them do so. If the sun shines for two weeks in a row, foragers can collect and the continuous stream of pollen and nectar brings the queen into top form. Then it suddenly rains for the next ten days. The honey factory has to slow from full steam to a standstill. The field bees are hanging around in the hive with nothing to do. The food flow is interrupted and the queen slows egg production down. Dissatisfaction arises and the swarm drive awakes.

Sometimes the drive arises without apparent reason because bees are bees and the drive to multiply is a natural instinct, as it is in all living organisms.

So bees are unpredictable during the spring swarm period. Beekeepers who do not want to be taken unawares by a swarm must check their colonies regularly for signs of swarm preparation, such as the appearance of queen cells (swarm cells) containing eggs, along the edges of brood combs. If these are present, so is the swarm drive. To prevent a swarm, all frames in the brood nest must be drawn out and any swarm cells broken off and removed. Bees will then not swarm for at least the next nine days because it takes this long for an egg in a swarm cell to be closed and a larva to metamorphose into a queen. And the old queen will also leave a hive that has no closed queen cells. Should a swarm cell be overlooked during an inspection, a swarm can be expected in the next couple of days and the beekeeper should get out a swarm-collecting basket.

A single inspection is not enough, because the swarm potential persists for three to four weeks and beekeepers must continue with their inspections and removal of swarm cells until no more are found. The drive is then suppressed, because during the three to four weeks that a colony prepares to swarm, occupied brood cells that initially led to the arousal of the swarm drive are now free again. In short, a fully occupied brood nest encourages swarming and for the next three weeks the beekeeper must remove all the swarm cells. The queen lays very few eggs, because the colony is preparing to swarm. Many young bees then emerge from their cells in this twenty-one-day period and free up brood space. The queen now has somewhere to lay eggs and the nurse bees have larvae to tend. The blocked potential of the colony that led to the drive has been removed and the drive suppressed.

The killjoy as saviour

Providing bees with more space and removing swarm cells is the traditional way to prevent swarming. There are other ways to either prevent or suppress the swarm drive when it surfaces, but to describe all these in detail is beyond the scope of this book.

We will, though, consider one aspect more carefully here, because it exemplifies the amazing abilities of bees as survival artists. After a few contests between newly emerged queens, members of the colony that remained in the old hive will then have a new queen – if she returns from her nuptial flight! While this is usually the case, it can happen that the queen never reappears. Perhaps a swallow catches her and offers her to its nestlings. Or she may lose contact with her escort and be unable to find her way home.

Losing a queen is not necessarily a disaster for a colony. The old queen continued to lay until she left. Should a queen die suddenly, perhaps squashed by a careless beekeeper, the colony is very soon aware of this, because there is no longer any queen pheromone and what was present soon disperses. The colony feels 'queenless', is disturbed and buzzes loudly and continuously.

In such an event, the bees resort to an emergency queen replacement. Worker bee larvae are normally fed with nurse's milk, which is similar to royal jelly, for just three days. With the loss of the queen, nurse bees select several larvae that are less than three days old and feed them with royal jelly instead of nurse's milk. By this means, in the case of a sudden loss of the queen, worker bee larvae can be converted into queens. You can tell this has occurred in a colony if mounds projecting from the surface at the edges of brood combs appear; beekeepers refer to these as queen replacement cells. The worker bees somewhat enlarge and elongate normal worker cells so that the larger queen has enough space to grow. A new queen emerges about twelve to fourteen days after the death of the old queen. No signals pass between the potential queens; any competitors still sitting in their replacement cells are quickly dispatched by the first queen to emerge; she opens their cells and stings them. Soon thereafter she sets off on her nuptial flight.

Bees remaining in the old hive whose new queen does not return from her nuptial flight do not have the option of a replacement queen because they need to have young larvae in the brood. Those larvae left behind by the old queen who left the hive with the swarm are already too old to be converted into queens by feeding them royal jelly. The colony would collapse unless a beekeeper came their aid. The queen pheromone is absent, order no longer reigns and anarchy breaks out. Nurse bees begin to feed other bees with royal jelly, these develop ovaries and soon begin to lay eggs. But worker bees are unfertilised and their unfertilised eggs hatch into drones. Worker bees who lay eggs are called drone mothers.

The surfaces of brood combs in normal colonies are flat and almost all the cells are closed. If a brood comb contains only unfertilised eggs laid by drone mothers then the surface of the brood comb is buckled because the drone mothers lay their eggs in worker bee cells that are too small for drones and the caps are domed to accommodate them. Without a queen or a new generation of workers, and with no harmony in the hive, the colony collapses. First it becomes smaller and weaker, and eventually neighbouring colonies notice. We will return to the consequences of this in the next chapter.

A beekeeper can save a colony whose queen does not return from her nuptial flight. Should a beekeeper find no young brood three or four weeks after the swarm has left the hive, they can do a 'queen cell test' to see if a queen is present or not. A frame containing a brood comb with very young larvae still being fed nurse's milk is taken from another colony. All the bees of the donor colony are brushed off the comb, before it is lowered into the brood nest of the hive to be tested. If a queen is present, the introduced brood will be fed as usual and the closed brood comb will have a flat surface. If there is no queen, the recipient bees immediately switch into queen replacement mode and the beekeeper would find the large queen replacement cells in the brood nest in which the colony is busily raising new queens from the young introduced larvae. Once a queen emerges, has killed off her competitors, undertakes and returns from her nuptial flight, everything is again in order. The colony will survive, thanks to the intervention of the beekeeper.

Rotten eggs in a healthy colony

After a colony loses its queen, worker bees eventually start to lay eggs. Worker bees in intact colonies also sometimes lay eggs, which would develop into drones if they were not first eaten.

A sociobiologist's explanation for workers laying and eating eggs is based on theories of 'selfish genes'[4] or kin selection.[5]

The underlying ideas are simple and ingenious, but the consequences are highly complex and would easily fill a book on their own.

It is simpler to consider the following scenario: if worker's eggs are artificially prevented from being eaten, only about a quarter of them survive, whereas almost all of those from the queen develop normally.[6] Within the confines of the hive, the inborn behavioural hygiene of worker bees obliges them to eat the eggs of their sisters and safeguard the queen's monopoly on motherhood.

Breeding queens: Manipulation of the smallest larvae

The ability of bees to convert worker larvae into queens is exploited by beekeepers in breeding new colonies. They begin by removing a queen from her colony and making it queenless. Soon the colony responds by constructing queen replacement cells, but the beekeeper does not allow these to mature. Nine days after the removal of the queen, all the replacement cells are taken from the hive and a 'breeding frame' placed in the queenless hive. This is a normal frame, except that instead of the usual comb foundation it has one or two cross members, each of which carry a row of queen cell bowls. A brood comb is now taken from a hive that has characteristics the beekeeper would like to cultivate, such as gentleness, slowness to swarm, and energetic foraging. The beekeeper searches the brood comb with a magnifying glass to find very small young larvae that are less than three days old. These are lifted out of their cells, together with some nurse's milk, with a micro-spoon and placed into the small bowl-shaped foundations of the artificial queen cells on the breeding frame.

The loaded breeding frame is then placed into the hive of the queenless recipients, where the workers immediately build up the queen cells, convert the foreign young larvae into queens by feeding them with royal jelly, and cap the cells after nine days. The beekeeper now encloses each queen cell in a small cage that also includes a number of workers. These are fed 'through the bars' of the cage by their sisters outside and have a special purpose. Twelve days after being spooned into the artificial queen cells of the breeding frame, the larvae metamorphose into queens, emerge from their cells and are received and tended by the workers in their cages.

Now it is the beekeeper's turn again. The queens are now individually persuaded to enter separate small cages so that they cannot kill each other and making it easier for the beekeeper to handle them. The queens from the breeding frame are now distributed among a number of small hive boxes that contain only two or three frames each. A single young queen is allowed to enter each of these small hives, which are transported some distance away from the old hive to prevent the field bees from returning to their old home. Several days after the move, the queen comes into season, and if all goes according to plan, four weeks later the first young bees emerge in the beekeeper's cultivated colony.

In this manner, beekeepers can prevent swarming and breed new young colonies. The possibility for apiarists to breed bees is important – and not just to raise highly productive bees. Selective breeding may also help address the challenges and dangers facing honey bees today, which we look at in the following chapters.

5

Bees as Aggressors

Imagine this scene: it's a summer Sunday at the end of July. The warm morning and bright sunshine morning invite one to take breakfast on the terrace. The table is laid and the aroma of coffee wafts in the air; it's time to read the paper and relax. A honey bee collects a small sample of honey from the open jar on the table and flies away, returning soon after with company. One or two sisters now also know about this rich source: not just nectar, but ripe honey that they can just take home and bring directly to the stores. Their colleagues need to be informed.

A telephone rings inside the house: it's the usual Sunday family conference. Returning to the breakfast table forty-five minutes later, one is faced with total chaos. The honey jar is surrounded by hundreds of bees, tumbling over each other to get to the honey. Squadrons of flying bees circle over the table and individuals dive down to land and run about searching the table, cutlery, plates and chairs. The air hums with their activity. The first three scouts told their colleagues about their discovery and a neighbouring hive have also got wind of it. It's time to plunder and the resulting tumult is an alarming sight. Although it's not really a cause for panic, what has happened here?

By the end of high summer, most flowers and trees no longer carry blossoms. Bee colonies that have reached their peak at this time of the year are frustrated. The honey factory has many highly motivated workers but no raw material. There is little nectar to be found and scout bees start to behave cunningly. In spring they would never notice an open honey jar on a breakfast table: their attention is drawn to the optical stimuli of flowers and the perfume of nectar, so the mild odour from an open honey jar is totally eclisped. In high summer, though, it's a treasure.

For the troubled soul at the breakfast table, the problem can be solved without calling the fire brigade. Gently shake the bees off the honey jar – no need to fear, they are busy searching and not out to sting – close the honey jar, clear the table and about an hour later the terrace will stand peacefully in the sunlight again without a trace of a bee.

Hostile takeovers: How bees steal from one another

Bee colonies whose honey stores are discovered by scout bees from another hive do not get off so lightly. In times of poor honey flow and when there is little nectar to be had from flowers or trees, scouts search for additional sources. They undertake a little industrial espionage to determine the strength of other honey factories in their neighbourhood. This is not particularly difficult. All scouts need to do is try to enter the factory door of another colony: simply alight and march with confidence towards the entry. The question is: are there enough guard bees to detect the intruder's foreign odour? Is collaboration in the hive effective enough that defensive behaviour can begin immediately? If the answer to both these questions is 'yes', the scout does not stand a chance. She will be approached, jostled, and if she does not promptly remove herself, will be fighting for her life.

But if the defence is only half-hearted? The scout bee then has a chance to slip through the guards and into the darkness of the foreign factory. Once inside, she is virtually out of danger. Very few bees inside the hive are in defensive mode and the scout can take her time to explore. She finds the honey store, collects a small sample and heads for home. There she informs her sisters of her find, and foragers fly off to the honey stores of the neighbour as if it were a blossoming orchard in spring.

Ever-increasing numbers of foragers, now robber bees, take off, and the colony under attack is defenceless. It is soon overwhelmed by the sheer numbers of attackers and its honey factory is doomed. Plundering foragers force their way into the hive and, although guard bees put up resistance here and there and some fighting takes place on the combs, there is no coordinated counterattack. Bit by bit, the resistance ebbs and the attacked join the attackers. Why risk one's life when the invaders appear to be such a well-organised and functioning community? The honey stores, and even open brood cells, are then trustingly cleared out and transferred to the plunderers' hive. Eventually the honey factory is emptied and the combs destroyed. A few dead guard bees and closed brood cells remain. The pupae in these will soon die from the cold.

The bees and the beast

How can a colony become so weak that it is not able to defend itself against thieving neighbours? Perhaps the queen did not return from her nuptial flight and the brood combs are buckled and filled with drones. There is a lack of harmony and motivation among the workers. Or it could be that the queen is old and unable to lay enough eggs to keep her colony strong. But perhaps for some time the colony has unwittingly hosted a malevolent enemy weakening it from within. This enemy has a name: *Varroa*, or *Varroa destructor* to be more precise. The *Varroa* mite.

A killer takes over the world

The *Varroa* mite has been a secretive and destructive parasite in all honey factories in Europe for a good forty years, and elsewhere for twenty or thirty (see below). It is an alien bloodsucker from Asia, where it developed in the last ice age together with the eastern, or Asian, honey bee, *Apis cerana*. Asian bees are no friends of these 2-millimetre-long beasts with their UFO-shaped bodies, but they have learnt to cope with them over an evolutionary time span of a thousand years or so. The mites live on bee haemolymph, or blood, and cannot survive without it. They hide themselves as best they can in the colony, but Asian bees seek them out, destroy them and a balance is struck in which the bees and the mites exist together as familiar enemies.

Western, or European, honey bees, *Apis mellifera*, were not infested with *Varroa* mites until late in the eighteenth century, and they would still be free of them had the mites been forced to find their own way west. Within a hive they can move remarkably rapidly, but to cross thousands of kilometres they needed some help. Of course they got this from humans. At the end of the nineteenth century, an enterprising officer in the army of the Russian tsar arranged to have some bee colonies from the Urals sent to him in east Kazakhstan to pursue his beekeeping activities. This was the beginning of a success story in which beekeeping prospered across the Urals, with western hives moving east. By the end of the nineteenth century they had reached Vladivostok and the native habitat of *Apis cerana* and the *Varroa* mites. Populations of western bees in the east also increased with the completion of the Trans-Siberian Railway in 1904. Migrants from Western Russia, seeking their fortunes in Siberia, took their bees with them and contact between the western and eastern bees became more frequent. However, *Varroa* mites did not immediately settle in western hives. Instead, not infrequently, a strong western colony invaded and plundered a weaker eastern colony and carried a few mites back to their own hives. *Varroa* mites do not all belong to the same species and many are highly specialised and only able to live with a single specific host: in this case, *Apis cerana*. Eventually, though, a type of *Varroa destructor* found conditions in which it could reproduce in *Apis mellifera* hives. Here, they found a paradise, because *Apis mellifera* had no conception of the destructive guests they had brought with them or how to deal with them.

During the 1960s, stories of unbelievable honey harvests in Russia began to spread and *Varroa* jumped west. Infected colonies bought by beekeepers in Siberia and imported across the Urals carried mites with them to Europe.

Mites took a different route to the American continent but were again aided by humans. Settlers from Europe in 1622 took honey bees with them to America where, until that time, this kind of social honey bee did not exist. The descendants of these bees reached Japan in the second half of the nineteenth century. At first nothing untoward happened and then, in 1958, the first *Varroa* mites were found on western honey bees. A Japanese beekeeper living in Paraguay in 1971 introduced colonies from his home country, together with *Varroa* mites, and so they jumped the Pacific. They were found in southern USA in 1987 and by the 1990s were distributed throughout America.

Varroa mites spread across the world in just fifty years, and present an enormous challenge to beekeepers. Australia remains the only country spared from the invasion. What is it that makes this intruder so dangerous?

The menace of **Varroa** *mites*

Varroa mites feed on the haemolymph, the equivalent of blood in bees. They cling onto the bees, bite into their bodies with their sharp mouthparts and access the bee's blood. Mites infest the brood because they can only reproduce within the closed pupal cells of workers and drones. A female mite slips into a larval cell just before workers cap it, enclosing the mite in the cell with the pupating bee. The female mite first lays an egg that develops into a male. This is smaller than the female, does not have an exoskeleton and will never leave the cell alive. The female mite then lays more eggs, all of which turn into females. These reach sexual maturity within the cell and are fertilised by the male. A worker cell that has been closed for about twelve days will host one or two mites, and that of a drone, which is closed for two days longer, will have up to three mites within it. When the young worker or drone emerges from the pupal cell, the female mite also leaves, with up to three daughters. The male mite stays behind and is eaten by cleaner bees.

After a few days, the mother and her young mature offspring each search for a brood cell about to be closed, enter it and repeat the cycle.

The problem arises because the western bees do not recognise a *Varroa* infestation. While eastern bees detect *Varroa* mites, perhaps from their odour, and clear them out, mites remain undetected in western beehives. A single female mite can produce up to nine offspring in three reproductive cycles, and these produce another nine each in three cycles. Thus, infestation of a western bee colony by a single mite can, after only nine reproductive cycles, lead to a mite population of over 700 individuals. The reproductive cycle of a mite is about three weeks long, meaning that a bee visiting a flower in May and picking up one mite left there by a colleague from an infested colony could end up with 700 mites in her own colony at the end of October.

The path to catastrophe is paved. Young mites in brood cells live on the blood of larvae and pupae, although these usually emerge alive from their cells, allowing the mites to escape. Bees emerging from infested cells are nevertheless weakened, their metamorphosis retarded, and they are often smaller and short-lived in comparison with their uninfested sisters. Mites also promote bacterial infection and, consequently, host pupae fall prey to pathological viruses and bacteria of every kind, from harmless ones to carriers of bee sicknesses. The uncontrolled spread of these illnesses or infections in pupae can severely cripple or even kill them.[1] Young infested bees often emerge with deformed wings or limbs and are not able to perform their assignments in the hive.

A bee colony can sustain an infestation in its first year. In October, 700 uninvited guests is a small number in a hive of 7000 to 10,000 bees, and the colony will survive the winter. At the end of March there will be 5000 bees and 300 sexually mature *Varroa* mites, because their numbers also decrease in winter. They cannot reproduce without the brood, and not all those carried on the bodies of bees survive.

The drama that can lead to the colony's downfall begins in spring. At first nothing is seriously wrong, and from April onwards the queen covers a large area of the brood nest. *Varroa* mites are included in many cells but enough young and healthy bees emerge from uninfested cells, although the number of mites rapidly increases. If there were 300 at the end of March, by the beginning of May there would already be 2000 mites infesting the colony, and almost 20,000 by the beginning of June. There is now a mite for every second bee, and many workers are physically impaired. On top of this, the colony normally reduces its brood production around the time of the summer solstice, and the brood area is smaller. Fewer open cells are available for *Varroa* mites to enter, although there are many more mites searching for them. Several female mites may now infest a single cell. The number of crippled bees increases, as does the pressure from mites on the remaining open brood cells. In July and August, when the colony needs to raise its winter bees, it is already too late. Virtually all the brood cells are now infested with mites and scarcely any unaffected winter bees emerge. The colony is overwhelmed by early autumn at the latest. Sick bees fly out of the hive and never return. Crippled bees drag themselves to the hive entrance, and neighbouring colonies attack the defenceless honey factory. The colony collapses.

Bees and humans against the mites: Strength in unity

The widespread infestation of bee colonies in Germany made it obvious that without support in the struggle against *Varroa* mites, western honey bees would not survive. Indeed, at the beginning of the 1980s, German beekeeping appeared to be at an end. Thirty per cent of the colonies were lost in winter and many apiarists gave up. How can one master such a situation?

There were effective poisons against mites available which could be used in bee colonies, and some of these chemicals were very successful in killing mites and apparently harmless to bees. But the use of chemicals in a hive raises two problems. The chemicals accumulate in wax and eventually appear in the honey. Honey containing pesticide? Not exactly an appealing prospect. The second problem is that wherever pesticides are applied, the targets all too often become resistant. Eventually a 'super mite' evolves, against which one can only use an even more toxic substance. So is there an anti-mite agent that does not accumulate in wax and that will not lead to a super mite?

Just such an agent is present in large quantities in forests, and birds have known for a very long time how to use this against irritating parasites. The substance is formic acid, produced by ants. Blackbirds have a curious behaviour around ant nests. They throw themselves flat onto the ground and flutter their wings to attract the attention of the ants. The ants feel attacked and respond by squirting the invader with formic acid. This is exactly what the blackbird wants, and it spreads the fumigating formic acid through its feathers. Any parasites that have settled there either die or leave, hoping to find another more acceptable host.

What works for birds works for bees

Formic acid vapour introduced into the hive causes *Varroa* mites to drop off the bees and leave the combs. Better still, the formic acid penetrates closed brood cells containing reproducing mites, but without harming larvae or pupae. So how does one introduce formic acid into the hive? Sending a horde of ants into every hive is not practicable.

Basically, the solution is not difficult, but it requires the beekeeper. A formic acid treatment is applied at the end of summer or start of autumn, when the honey harvest is over. Sixty per cent formic acid is introduced into the hive with an evaporator, which can be a simple sponge. This is soaked in the formic acid solution, the lid of the hive raised, the sponge placed inside the hive and the lid closed. Formic acid from the sponge vaporises immediately if the ambient temperature is high enough, and in a strong colony the hive is warm. Slowly the acid fumes fill the hive. The bees are not entirely pleased and avoid the unpleasant sponge, although in general the slightly acid atmosphere in the hive is harmless. A sheet of aluminium foil laid under the hive collects the falling mites. Acid in the sponge dries out after a few days and the procedure is then repeated. An overwintering colony occupying two cases must be treated four times over a period of about three weeks, by which time 250 millilitres of formic acid will have evaporated.

The colony is then virtually free of mites and can raise enough healthy winter bees to survive the winter.

The battle begins again in spring. Our colony may have robbed a mite-infested neighbour in late autumn and one of the robbers brought back to her hive not only stolen honey but also a *Varroa* mite. This mite would begin to reproduce as soon as the new brood were established and in two months would have raised an imposing number of offspring. The hive is again under pressure from *Varroa* mites. Now what? Formic acid is no longer an option, because this would taint the honey. One strategy for beekeepers and their bees is to exploit a behavioural tendency of the mite itself to bring about its downfall. Mites reproduce in closed cells, and because drone cells remain closed longer than worker cells, more mites have time to develop within them. *Varroa* females have adapted through evolutionary selection to discriminate between drone and worker cells, perhaps by their different odours. *Varroa* females prefer drone cells and whenever possible let themselves be enclosed in these instead of worker cells.

Bees build many drone cells in spring because they will need drones in the coming swarm period. A beekeeper who now hangs an empty frame in a strong hive will have, after less than a week, a frame containing only drone cells. The queen lays only drone eggs in these cells and, yes, *Varroa* mites establish themselves in these and develop along with the drones in their cells. As soon as the cells are capped, the beekeeper removes the frame from the hive, cuts the infested combs out and melts them down. A fitting end for the mites. The empty comb is returned to the hive and the process continued until the end of June, after which bees no longer build drone cells.

Varroa mites remain a problem for beekeepers everywhere but they do not pose an overwhelming threat as long as apiarists support their bees in their struggle against mites. With the help of decoy combs in summer and formic acid treatments in autumn, beekeepers and western bees can achieve what eastern bees have through evolution, and that is to keep the infestation small. Bees will never rid themselves of mites but can tolerate them in small numbers.

The Struggle for Survival

Early humans would soon have found the treasure horded by honey bees. Hunters and gatherers of the savannah were rewarded with a feast if they discovered a honey bee nest. Honey must have seemed like a gift from the gods, possessing an unearthly sweetness and created by beings whose abilities and artistry they did not understand. Perhaps this is the source of our fascination with bees. From the very beginning, they offered something unique and made humans happy.

Today, nearly two million years later, we perhaps appreciate this uniqueness more clearly and intensively. The ingenuity of early apiarists brought bees out of the wilderness to live closer to people, making it possible to develop the first simple beekeeping practices. Natural science and research over the last 300 years or so has significantly deepened our understanding of the bee colony superorganism and laid the foundation for modern apiarist practices, namely honey factories. Accelerating globalisation in recent times has resulted in people and bees coming even closer together. For bees, this means that without the help of humans there are many areas in the world where they could not survive *Varroa* infestation. For humans, the lack of bee pollinators means they would face an even greater challenge than that already confronting them in their attempts to provide enough nourishment for an expanding world population.

The human race would not necessarily starve without bees. Cereals in particular and many other food plants are wind-pollinated and do not need insects. Vitamin requirements of humans are another matter because most vitamin C in our diet is obtained from plants pollinated by insects and 80 per cent of the indigenous fruit trees in Germany are dependent on honey bees. Without bees, there would not be much hope for cherries, apples, plums or pears. Agricultural production dependent on beekeeping exceeds that of poultry and makes bees, after cattle and pigs, the third most economically important domestic animal.

No wonder there is anxiety when stories of high worldwide honey bee mortality circulate in the media. What, exactly, is the state of affairs?

The myth of the dying honey bees

To begin with, statistics tell a very clear story. The number of bee colonies in Germany has decreased continually over the last few decades. Members of the German apiarists' society (Deutsche Imkerbundes, or DIB) maintained over 1,000,000 bee colonies in 1977. By 2008 this had dwindled to less than 700,000. Thirty per cent of all bee colonies had disappeared within thirty years. But what do these figures really tell us? Is this evidence of bees dying off or is there another reason? The question also arises when one considers the subsequent development of beekeeping. Since 2008, the number of colonies managed by DIB members remained constant, or even slightly increased. If bees are dying, why has the negative trend not continued?

Taking into account both the number of beekeepers belonging to the DIB and the average number of hives in their care, it would appear that the phenomenon has nothing to do with bee extinction. In 1971, the DIB had about 900,000 members. These slowly decreased, as did the number of bee colonies, until 2008. With the reunification of West and East Germany, more apiarists joined the DIB, membership increased and then again subsided. In 2015, with over 100,000 members, the DIB was larger than in 1971, but whereas a beekeeper in 1971 managed an average of eleven colonies, forty-five years later this had shrunk to seven – a decrease of about 36 per cent.

These figures do not support the notion that honey bees are dying out or that beekeeping is vanishing. At present, the number of beekeepers in Germany is increasing at about 3 to 5 per cent per year, and the number of colonies is also growing. What the statistics reveal is the consequence not of a bee die-off, but of a basic change in the structure of beekeeping. This structural change is related to growth in salaries and income, the cost of food and how much people are prepared to spend on it. The change began after World War II and with the economic success of West Germany.

Changes in beekeeping: From a business sideline to a hobby

In 1956, about 13,000 tons of honey were harvested in Germany from around 1.3 million beehives, an average yield of 10 kilograms per colony. The price of honey at this time, converted to the present equivalent, was about €3.20 per kilogram. A beekeeper managed an average of fifteen colonies and usually kept them together in a bee house, not in magazines. With a harvest of 150 kilograms of honey per year, beekeepers could reckon on a capital gain of €480 from their bees. Considering that the average annual earnings of a worker in 1956 lay at about €2600, beekeepers could, in the 1950s, subsidise their yearly income by more than 18 per cent.

Only fifteen years later, in 1971, the average apiarist now had a yield of 167 kilograms of honey, for which, with slightly increased prices, they could get €690. Meanwhile, the average annual earnings of a worker had increased to €7700. The yearly income from beekeeping could subsidise a beekeeper's salary by only about 9 per cent. In 1975, the average yearly income was now over €11,000, while honey prices remained stable and the honey harvest that year was poor. The subsidy from bees diminished to only 4 per cent of the annual wage.

The trend is clear. In 1956, beekeeping in Germany was a profitable sideline. It contributed an additional 18 per cent to the annual wage for beekeepers with fixed locations, and almost double this for those who travelled and could expect to harvest up to 35 kilograms per hive. Twenty years later, beekeeping no longer made a profit and was just extra work. In the 1980s, *Varroa* mites arrived and many apiarists gave up or could not find anyone to take over their small businesses. The number of beekeepers and colonies sank.

Simultaneously, another development took place. From the middle of the 1970s, after the Club of Rome published their report *The Limits of Growth* (1972), a new philosophy emerged concerned with the relationship between humankind and the environment. Environmental protection became a priority and the ecological movement was founded. The first 'Green' political party was founded in the 1980s. These changes and the associated shift in perceptions of nature and the environment induced some people to take up beekeeping. German beekeepers also changed their practices, adopting magazine hives – our modern honey factories – with the result that beekeeping once again became more profitable. The average yield per colony in 2009 was about 24 kilograms and the price of honey €8.90 per kilogram. Beekeepers with fifteen hives in 2009 could expect to earn €3200 from their bees. Their yearly income was now €41,000, so that the profit from their hobby had now risen to 8 per cent of their annual income, almost twice that in 1976.

Over the last forty years, beekeeping in Germany has evolved from being a good agricultural sideline into a profitable hobby. Beekeeping enjoys growing popularity in the context of changing attitudes towards nature, and there is no danger at present that honey bees in Germany will become extinct.

The same is true in other parts of the world. Statistics from the nutritional and agricultural organisations of the United Nations indicate that the number of human-managed beehives had significantly increased worldwide in the last fifteen years. The number of honey bees is increasing in Africa, South America and even in China, despite their problems with pollution. The worldwide demise of *Apis mellifera* appears unlikely.

A threatened life: Why bees have it hard

So, then, it would seem that everything is rosy. Unfortunately not, for although the worldwide death of honey bees is not to be feared at present, the situation of their close relatives is not so positive. Besides the honey bee, there are about 560 different species of wild bees in Germany alone. These are all 'solitary' bees, which unlike honey bees and bumblebees do not live in colonies. These 'single bees' are highly specialised and need very specific locations and conditions, and the availability of specific plants or other insects to nourish their brood. Solitary bees find increasingly fewer possibilities as the landscape loses its structure and easy-care home gardens become ever more monotonous. Thirty-five per cent of these species are threatened with extinction. Solitary bees live precariously.

Honey bees are also not living in paradise. It is already obvious how heavily dependent they are on humans to survive *Varroa* infections. Humans also put bees under severe pressure by arranging the bees' world to suit human needs. Almost the entire living space for insects and animals in Germany is used by people, and only a small portion of the Federal Republic is set aside as nature reserves. The rest is given over to industry, roads, residential areas, agricultural lands, forests or green recreational areas. This has consequences for honey bees. Not long ago, a bee house was a part of every farm. Honey bees were economically important and kept in villages surrounded by fields and meadows as a profitable business sideline. Today, agricultural areas are merely green deserts for honey bees. Previously, dandelions bloomed in the meadows and sometimes – to the irritation of farmers – so strongly that meadows looked like large yellow handkerchiefs strewn across the landscape. Meadows, when allowed to grow to make hay, are filled with summer flowers in June and July. Here, bees can find a rich source of nectar and a diverse and nourishing selection of pollen.

Now the meadows grow only grass. Large dairy farms and, more recently, biogas companies need food and raw material. To produce enough for their businesses to survive, farmers resort to planting grass types that can yield up to six cuttings per year. No flowers bloom in fields that are mowed six times a year. There is nothing here for bees. This is not a criticism of modern agriculture, under pressure in a globalised world, just a statement of fact. Agricultural areas are so efficiently and intensively used that frequently they no longer offer an adequate living space for insects and bees. Bees in the country have a far more difficult life than those in the cities! Beekeepers in country areas, when they do not travel with their bees to follow the honey flow, often have significantly smaller honey harvests than their colleagues that keep their bees near towns.

The reason is clear. Bees find what they need in and near towns, because in general towns are greener and have far more diverse flora than the countryside. Parks, residential areas with old fruit trees in their gardens, cemeteries, abandoned gullies or fallow fields and, not least of all, trees along the streets all offer bees a rich, diverse and year-round source of nectar and pollen. A beekeeper in the Hamburg region in 2015 could harvest, on average, 37 kilograms of honey from each colony. The harvest of beekeepers in the Weser-Ems region, a predominantly agricultural area, averaged significantly less – only 30 kilograms per colony.

Extinction remains a possibility

The Swiss documentary film *More than Honey* from Markus Imhoof dramatically reveals the extreme abuses of nature beekeepers go to in order to satisfy human economic demands. Imhoof introduces us to an American apiarist who, not interested in harvesting honey, earns his income instead from 'pollination' beekeeping. The goal of this practice is to bring as many bees as possible to cultivated orchards in bloom to generate optimal pollination and fruit yield. The strategy has been practised for many years in Germany, for example in the Old Lands around Hamburg. Bee colonies are brought into orchards at the beginning of cherry blossoming, and, when the weather is good, ensure that cherry, apple and pear trees all bear heavily. The fruit grower has a good harvest, the beekeeper a fine spring honey from the fruit trees and usually an adequate financial reward from the grower for the service. But there are no industrial pollinating concerns in Germany operating on the scale of those in America.

Industrialised bee colonies: Pollination beekeeping in America

American pollination beekeepers are not particularly interested in honey. Their main aim is to provide a paid pollinating service in which they bring as many bees as they can to sequential honey flows and receive a weekly payment for each colony. The beekeepers let their bees overwinter in the warm Florida climate, which saves on feeding them. In February, the colonies are loaded onto large trucks and taken 4000 kilometres across the continent to California and the largest almond orchards in the world, a gigantic monoculture where more than half the entire bee colonies in America are concentrated. In March, the colonies are again loaded onto their trucks and moved 1000 kilometres north to pollinate the apple orchards in Washington, and in May, 2000 kilometres east to fields of canola and sunflowers. In June another 2500 kilometres east to blueberries in Maine. Then 1000 kilometres south to pumpkins in Pennsylvania, and in August, 1500 kilometres back to Florida. A bee colony travels 12,000 kilometres each year, is located primarily in monocultures and nourished on a diet with very little diversity. The brood combs are also frequently removed to prevent swarming and used to artificially found new colonies that can then be employed to increase the pollination premium. With frequent changes of location over many thousands of kilometres, the associated climate change, absence of pollen diversity and loss of hive harmony through the frequent splitting of the colonies, it is no wonder that such a colony has had enough by the end of a season.

The first reports of massive unexplained bee deaths came from America. The malady is known as colony collapse disorder (CCD), and some pollination beekeepers lose up to 30 per cent of their colonies each winter. Hives that were filled with bees in October were found to be empty in December when they were opened for feeding and to prepare them for the almond blossoming. An unexplained phenomenon?

Beekeepers in Germany have also observed similar events, although they were not pollinating apiarists. A colony appears strong and vital after feeding in August and ready for winter. It can then become so cold that bees gather into a winter cluster, and curiously, when examined, the cluster appears smaller than expected. In December, it is warm and then cold again, and the hive is suddenly empty. Not a single bee is to be found. What has happened here?

The colony may have been under severe pressure from mites and the formic acid treatment in late summer was not effective. Perhaps it was too cold for formic acid to evaporate or relative humidity in the hive was too high. Too many mites survived and most of the winter bees were crippled, so although the number of bees was high they were not fit. Most were already disabled when they emerged, and disease spread through the colony because the weakened immune systems of the bees provided no defence. Viruses tax the energy of bees; destructive bacteria and fungi add to the stress. Sick bees always leave the hive. Slowly, on frost-free days, the bees leave the hive until there are virtually none left. Noticeably high regional losses continue to occur in winter in Germany, but no unexplained CCD.

The massive death of pollination colonies in America is perhaps not really so mysterious. Everything points to the deaths being a consequence of procedural errors. An excessively operated pollination apiary works against the nature of the colony, not with it. A bee colony is an integrated living organism that needs seasonal change, a climatically stable location and a diverse and seasonally changing pollen and nectar supply. It also requires a harmoniously organised inner structure, with old and young bees, drones, brood, growth in spring and a decrease in numbers in late summer. A bee colony remains vital and can survive only if this seasonal rhythm is maintained. Repeated relocation in different climatic zones, an undifferentiated diet in various monocultures and the brutal splitting of colonies without regard for season or colony structure can only lead to weakening and the death of a colony.

Humankind as a danger to bee colonies

Poor management of bee colonies is not the only cause of their collapse. Human impact on the environment and the spread of pesticides in particular represent incalculable risks for the future of bees.

This became very clear in the spring of 2008, when reports of widespread bee mortality appeared on the front pages of German national newspapers. Thousands of bee colonies in an area between Rastatt and Lörrach in the southern Rhine region had died within a few weeks. Instead of approaching their summer peak and bringing in spring honey, they were dying.

An attempt by farmers in South Baden to protect their maize against the western root borer was responsible for the drama. The larvae of these beetles damage the roots of maize plants so severely that they cannot produce corn. To prevent this, seed corn was coated with a layer containing the insecticide clothianidin. Seeds treated in this way can germinate in the ground and are not attacked by borers. But the treated seed has to be sown and should the poisonous coating of the seed become free and not then bound or captured by the sowing machine, the clothianidin finds its way as airborne dust to flowers or the blossoms of fruit trees. This is precisely what happened in Baden. Unprofessional sowing of treated maize seed led to the spread of clothianidin dust, bees took this up from the flowers, and entire colonies were wiped out.

Bees are not only endangered by synthetic poisons. The list of honey bee diseases is longer than for any other insect species so far investigated. Nosema, chalkbrood or foulbrood can severely stress a colony or even kill it. Most of these diseases are caused by bacteria that find optimal conditions to multiply in the warm environment of the hive. Considering that bees live in continuous close contact with one another and in many stages of their development do not possess very effective immune systems, infections would be expected to spread quickly. From a biological perspective, it is remarkable that bees exist at all. The pupae have little immunity against bacteria, whether these are pathological or not. Every kind of bacterium that finds its way into a pupal cell can multiply without hindrance.[1] Normally such infections do not arise because the pupal cells are hermetically sealed, unless a *Varroa* mite is enclosed in the cell and brings viruses and bacteria with it, infecting the pupa when biting it. Molecular genetic analyses of the bee genome have shown that of all the insects investigated, bees have the fewest genes responsible for the immune system. So how do bees manage to stay healthy?

The answer to this question lies with the singular social structure of a bee colony. A bee colony is not simply a collection of 50,000 individuals, like a flock of sheep. Instead, the many single bees together form a superorganism that as a whole undertakes

THE STRUGGLE FOR SURVIVAL

assignments and achieves objectives that a single bee alone cannot. Comb building and brood rearing are part of the defence against disease. The bee colony has, in addition to the defensive possibilities of the single bees, a sort of social immune system functioning on the basis of their all living together. Bees within the hive are continually busy with cleaning cells and frames and also keep each other clean. Bacteria and fungus spores don't have much of a chance of spreading and endangering the occupants of a strong colony. Propolis is also employed to completely enclose anything that can rot or cause harm.[2] And there is no bee hospital. Sick bees are not tolerated by their healthy sisters. If they do not leave the hive on their own, they are attacked, forced out the door and not allowed back in. Bees apparently detect the odour of an infected colleague, and sick foragers are usually unable to find their way home. Bee colonies appear to reject their sick members in much the same way that our own bodies react to damaged or infected cells.

There are essentially only four threats to the wellbeing of bee colonies. The first is insufficient size. Small numbers of bees result in weak superorganisms that are vulnerable to disease. Beekeepers do their best to raise the strongest possible colonies to manage the second hazard, which is hive hygiene. Regardless of how large a hive becomes, unless the beekeeper regularly provides them with new combs, bees cannot keep the old brood and honeycombs clean forever. The old, dark and discoloured brood combs become a dangerous breeding ground for bacteria and fungi.

The third threat that has existed over the last fifty years in Europe and Northern America is the *Varroa* mite, described in detail in Chapter 5.

The fourth is the presence of pesticides introduced into the environment by humans. The possible consequences have been described at the beginning of this chapter. One matter is clear in relation to this hazard. In a world in which highly developed and intensively practised agriculture is required to nourish a growing human population, it is not possible to manage entirely without the use of chemical preservatives, herbicides and insecticides. But in order to apply these substances intelligently, it is critically important to know what effect they can have on bee colonies. There are many research institutes across the world that are occupied with the problem of protecting bees from these dangers, and they face a Herculean task. There is a clear recognition that the question 'does a certain chemical substance kill bees or not?' is far too simplistically formulated in order to understand the larger problem of the influence of synthetic chemicals on the environment and judge the consequence of this for bees. Approaching the wellbeing of honey bees from a scientific point of view quickly leads to frustration because of the nature of the organism under investigation. This is not the individual honey bee but the bee colony, an integrated unit with a multifaceted inner organisation involving complex interactions between its members. A simple example is provided by the negative effect of cold on bees. A single bee becomes stiff when the temperature falls to 10°C and dies if the temperature sinks below about 4°C. A bee colony placed in a cold room can survive temperatures as low as minus 40°C, provided they have enough honey to generate warmth. The threat of cold therefore has an entirely different meaning dependent on whether an individual or an entire colony is concerned.

The same applies to threats from human factors. Not every environmental danger for a single bee is necessarily a problem for an entire colony. On the other hand, a colony does not consist of only adult individuals. There are eggs, larvae in different stages and pupae. In their lifetimes, worker bees pass through several stages as summer or winter bees, in which they react differently to environmental conditions and also have very different levels of resistance to disease.[3] What is harmless for nurse bees can kill field bees. What is of little concern for adult bees can kill larvae. Queens and drones also have their own unique properties and characteristics.

Interactions between different chemicals introduces another dimension. Spraying a single substance on its own onto a flower may not harm a bee, but the poisonous cocktails that are not infrequently applied can be fatal.

Finally, there is the difficult problem of 'sublethal' effects. Not every poison kills immediately, yet the life of a bee can be so impaired that it dies eventually. The death of a bee colony in winter can occur naturally. However, in some cases it may be caused by a poison with an effect on the colony so subtle that it is not possible to determine its point of action.

The intense discussion over the use of neonicotinoids is a good example of how complex the matter can become. This relatively new group of chemicals, to which clothianidin belongs, is useful and highly effective from an agricultural point of view. But there are increasing numbers of reports that neonicotinoids have serious sublethal effects on honey bees. Such studies implicate these chemicals in the impairment of a bee's ability to orientate.[4] The discussions over this are by no means finished, and the interests of agricultural societies, chemical industries and nature conservation agencies are heavily in conflict. Discussion should be welcomed, for only through open and forthright debate based on scientific evidence will it be possible to continue with what has prevailed so far – the survival of bees in an environment characterised by industrial agriculture. Diverse research in this area continues to produce very surprising results.

Honey bees: Future caterpillar frighteners?

The caterpillars of large noctuid moths can cause a lot of damage when they feed on the leaves of cabbages, lettuces and other plants. The protein-rich caterpillars are themselves prey to a specific wasp and have developed an early warning system against this particular menace. Fine hairs on the caterpillar's body are tuned to the wingbeat frequency of the wasp and when it approaches, these hairs vibrate and caterpillars react by dropping immediately off the plant onto the ground, where they lie motionless and safe from the wasp, who hunts only moving prey.

Honey bees have a body size and wingbeat frequency similar to the wasps, and the caterpillar's simple sensory hairs are not able to distinguish between the two. Could bees be employed as a non-toxic biological deterrent to caterpillars?

In a simple experiment to test this, two large cages were set up in which paprika and soya plants were grown.[5] Caterpillars of a noctuid moth known to growers as a severe pest were placed in both cages. In one cage, the caterpillars were allowed to feed undisturbed. In the other cage, bees were allowed to enter to visit a feeder inside the cage. Flying bees disturbed the caterpillars so severely that they ate only about one-third of that eaten by the undisturbed individuals. It's possible to imagine that cabbage fields in the future might have interspersed rows of wildflowers to attract honey bees and discourage caterpillars.

Back to the future: Rediscovering old practices

Having grasped and understood that a bee colony is a single, integrated, living organism to be worked with, not against, it is not only industrial pollination apiaries that we are critical of. Intensive employment of bees in magazine hives, our honey factories, is also not beyond reproach. Does this kind of beekeeping, which ignores the needs and nature of a living organism, border on violation? Perspectives about the treatment of bees have changed, and the motives and goals of beekeeping have altered since its transformation from a business to a hobby. For some beekeepers today, honey yield is of less importance. Instead, their interest is in examining the practicality of extensive apiaries. What is the most suitable hive? How should one manage such a concern? There are some who do not want to harvest honey at all; they just want to have a colony in their garden to watch and enjoy. How should one care for such a colony to ensure that it will not collapse from a *Varroa* infestation? Questions have also been raised about keeping bees in a fashion that conforms to their specific needs and natures.

The words 'natural' and 'keeping' stand in contradiction to each other. It is not natural for any wild animal to be kept by humans, highly considerate though this care may be. There can be no talk of 'natural' the moment we no longer take animals and their products directly out of the wild state for our own use. The animals then usually do not survive. Plunderers of a wild bee colony's honey reserves take what nature offers and leave the robbed colony to their fate. Is this really what we want? A colony without honey reserves will, at least in European latitudes, most likely die in winter. The colony would survive longer if it was left in peace with its honey, or if one took its honey then fed it. And then we are back to unnatural beekeeping.

At what stage does such care for a particular organism conform to its specific needs and nature? The debates about the conditions under which poultry, pigs or cattle are kept are well known. How much space does a hen need? Can it scratch about or not? How many pigs can be kept in a sty before the stress becomes so great that they begin to chew on one another? The question receives much research and causes even more arguments. Attempts are being made to reconcile the interests of humankind with the wellbeing of economically important animals. Pig sties with wallows, runs, places to rest and exercise are certainly a more pleasant sight than cramped feedlots, and one has the impression that the pigs are a good deal happier too.

But what constitutes the 'specific needs and natures' of bees? This question has been loudly and vigorously debated over the last ten years. Does keeping bees in magazine hives, or honey factories, for the purpose of harvesting their honey conform to their specific needs and natures? The bees are not able to tell us, and the perception of their needs and natures is determined instead by beekeepers. This view is compared with reality and in extreme cases leads to the conclusion that nothing we do is in accord with the needs and natures of bees.

Some people express the view that bees should live in natural hollows and that closed, artificially constructed beehives are geometrically inappropriate. Such sweeping statements are extreme and probably incorrect. Bees are highly adaptable and thrive in many different kinds and shapes of hollows that they choose themselves. They do very well in magazine hives that are well managed by apiarists, and if this form of beekeeping contains stress factors then they should be identified. Until that is done, the question remains open.

Beekeepers are criticised for attempting to stop bees from swarming. Bees have a natural drive to swarm and multiply their colonies, while the management of honey factories prevents swarming to encourage the development of stronger colonies. The alternative? Let them swarm? Certainly, it is exciting to watch a swarm and to show your neighbour how it is captured, provided it occurs only once or twice a year. However, when a beekeeper regularly walks through their neighbour's garden to fetch their errant workers home, peace and beekeeping in residential areas are soon both at an end! Furthermore, we live in a cultivated landscape, not a forest. Where can a swarm that is not captured go? There are practically no naturally occurring refuges for swarms available. Often, the only trace left of a swarm a beekeeper has lost is a single comb hanging from a tree branch. The swarm was unable to find a suitable home site and in desperation began to build combs where they were stranded, until the first heavy rain washed them away. And even if a swarm found a hollow – perhaps an empty compost drum – to settle in, would they survive a mite infestation?

Caution is advised when proposing an idealistic view of 'natural' beekeeping, which has little to do with the reality of the world in which bees live, and have to live, today. This world is shaped and modified by humankind and what we have wrought and will continue to wreak, is not always positive. A 'back to the beginning' for beekeeping is unfortunately not practical.

Nevertheless, an awakened sensitivity to the original lifestyle of the bee colony and the rediscovery of wild bees may reveal a way to construct honey factories of the future that at the moment we cannot imagine. Thorough study of free-living bee colonies and open-minded examination of the present treatment of kept colonies may show how we could more fully appreciate their needs and better satisfy their requirements. Initiatives along these lines have already led to some interesting questions.

Bees and forests: Learning from their origins

The bees that apiarists keep today are domesticated livestock. They belong to the genus *Apis*, which has been cultivated by humankind and adapted to our agricultural economies. What happens to bees left in their natural environment and virtually uninfluenced by humans? Would such a feral colony, like the Asian species, finally learn through natural selection to cope with *Varroa* without help? And would these feral colonies possess a gene pool that could improve the vitality of all bee races?

Such questions not only lead to consideration of special breeding techniques but have also awakened an interest in a very old and original beekeeping practice, the 'Zeidler' method. The word 'Zeidler' comes from the Latin word *excidere* or 'excision' and refers to harvesting honey by climbing trees containing wild honey bee nests and cutting the combs out of the hollows. The practice goes a step further, however, including carving hollows in living tree trunks at a convenient height and fitting the hollows with covers and an entry. Zeidler beekeepers can keep and manage a large number of tree hives over a widespread forest area. The original Zeidler beekeepers organised themselves into a guild with regional privileges and even their own jurisdiction.

This form of beekeeping occurs today in a few regions in eastern Europe. The practice has been followed continuously in Baschkirien in the south Urals and was brought within the context of a World Wide Fund for Nature (WWF) project to Poland, where now more than 100 tree hives are scattered throughout the country and used for Zeidler beekeeping. This original method may not necessarily appeal to all beekeepers, but has attracted interest in Germany and could make a contribution to restoring an important component of forest ecosystems. A scientifically conducted Zeidler apiary could also provide us with an opportunity to look inside such a colony, which has so far remained unexplored. Honey bees are, after all, primarily forest insects and have adapted over their long evolutionary development to spending much of their lives within hollow trees and also to sharing their hollows with other organisms. We know virtually nothing about these commensals because we do not know what the inside of a tree hive looks like nor what happens there. A small ecosystem exists in tree hollows, the complexity of which we have only the most basic notion. Organisms that share a hollow with honey bees lead lives closely interrelated with those of the bees. Wax moths are an example. Wax moths lay their eggs in old combs and their caterpillars feed on remains of pupal cases left behind by emerging bees, and on pollen. The caterpillars eat their way back and forth through the combs and cover the walls of their tunnels with a fine silk web. A comb that has been occupied by wax moth caterpillars presents a rather disgusting chaos of webs, wax fragments, caterpillar faeces and wax moth pupal cases. Beekeepers are understandably offended by such a sight.

What beekeepers reject in honey factories is a blessing for wild bees. A natural forest-dwelling bee population multiplies each year through swarming. New home sites for swarming bees become available when new hollows form in trees or old hollows are freed by the death of a colony. New hollows do not occur all that often without the help of the Zeidler apiarists and abandoned nests are often filled with old combs that the new swarm cannot use. The wax moths play their role here. Attracted by the odour of the old combs they lay their eggs in abandoned nests and the caterpillars make way for new combs by destroying the old.

We may discover cooperation between bees and other organisms in tree hives, inhabited as they are by bacteria, fungi and a host of arthropods and other small creatures. To what extent these are involved with the lives of bees is virtually uninvestigated.

Some initial approaches in this area have been undertaken and a small arachnid has aroused the interest of practical beekeepers.[6] Book scorpions (pseudoscorpions), with their eight legs and long pincers, look like miniature scorpions: hence, their popular name. They are also predators like their large sting-bearing namesakes. Only a few millimetres long, book scorpions live on even smaller creatures that they catch in their pincers and kill

THE STRUGGLE FOR SURVIVAL

with poison from their mandibles. The prey is then sucked dry, as is the way of arachnids. Book scorpions can be found in bee colonies living in trees and also in colonies kept in old-fashioned straw hives. They were highly valued by a past generation of beekeepers for helping to free the bees from a parasitic tracheal mite that infested the bee's respiratory system. Modern beekeeping practice does not encourage book scorpions. Magazine hives are precisely constructed and provide no cracks or crevices to accommodate book scorpions, and the use of formic acid against *Varroa* mites is not good for them either. Nevertheless, the return of book scorpions to beekeeper-managed bee colonies could be helpful in supporting bees in their struggle against *Varroa* mites. One of these small predators is able to destroy up to ten mites a day. A careful study of bees living in forests and Zeidler tree hives could lead to many new insights.

Such investigations may provide answers to other questions, for example the effect of smoke on bees. There is perhaps no better known or more typical beekeeper's utensil than the smoker. Even those who have little or no interest in bees know that beekeepers blow a little smoke over their beehives to calm them down before working with them. This is a misunderstanding.

Smoke does not calm bees; on the contrary, it alarms them. Forest fires for a forest-living bee are a serious threat. Where there is smoke there is also fire and as soon as bees detect smoke they search for open honey cells in the hive and fill their stomachs from the stored honey. This is what the beekeeper intends, because after a puff of smoke bees retreat down between the combs and once full of honey they become less aggressive. The smoker strategy only works if there are enough open honey cells. Late in autumn, when most winter reserves are in closed cells, an attempt to 'calm' bees with a puff of smoke results in an unpleasant surprise. They attack and clearly demonstrate just how seriously smoke disturbs them.

Why then do they retreat down into the hive earlier in the year? An explanation in the literature is that bees are preparing for an escape by engorging themselves with honey. Alternatively, it is known that strange bees appearing at a hive entrance can gain admission by 'bribing' guard bees with honey. Are smoked bees hoping to find a refuge in a neighbouring hive? Apart from the fact that this would imply planning of a kind that we would not expect from individual bees, where would bees get the idea to leave their queen, who is certainly unable to fly away with them? We have to keep in mind that individual bees are like 'cells' in the 'body' of the colony superorganism. It is unlikely that the fate of the entire hive is dependent on the actions of single individuals the moment they smell smoke. There is also no reported case of an entire bee colony leaving its hive because of fire. Instead, there are many accounts of the tragic incineration of bee colonies when hives set out in forests are overcome by fire.

The behaviour of bees near smoke has other possible physical grounds. It would be interesting to investigate the thermo-physical effects on bees that had full stomachs. Such bees may heat up more slowly due to the increased body surface and volume caused by honey swelling their abdomens. They could then hope to survive a fire passing rapidly through a forest at tree height, where the temperature is lower than on the ground. Research with Zeidel tree hives could provide some information to support such speculation.

Can bees not see the forest for the trees?

We may also learn a little more from Zeidler forests about how bees find their way when foraging in thickly wooded areas and how, under these original living conditions, bees developed all that our present research has discovered. Research on the orientation behaviour of bees – their ability to fly to flowers, gather nourishment and find their way home – has typically been undertaken in open country. This is not an oversight, because the environment of bees nowadays is usually not in forests but in the open, which is convenient for researchers because they can identify possible optical landmarks. But can we completely understand the orientation abilities of bees when we conduct research only under these conditions?

It is certainly not absurd to suppose that a very different situation prevails in thick vegetation, which presents a different set of challenges for orienting bees. Studies on recruitment by scouts and foragers to a feeder set in thick forest show that fewer new arrivals appear over time than when the feeder is located in an open field. Orientation to feeders in forests is apparently more difficult for bees than in the open country where they can follow a flight path guided by a relatively small number of separate visual landmarks. There are so many landmarks in a forest that one may ask how the bees find their way at all.

They have a solution to the problem. Observing the arrival of experienced foragers, who have danced in the hive, at a feeding site in a thick forest reveals some interesting behavioural traits. First, bees fly through the trees and not over them, even if the distance from the hive to the feeder is several hundred metres. Experienced bees arriving at the site all have extruded Nasonov scent glands and carry out extensive buzzing flights. When observing from a position close to the feeder, so that oncoming bees can be seen through the trees, it becomes clear that new recruits arrive not more than ten seconds behind experienced bees and most of them as though attached to the experienced bee by a thread. Should the leading bee fly to the right of a tree, the follower usually also flies to the right. When a leader flies to the left, so does the follower. The last stretch of the flight, in full view of the observer, takes place in most cases at a height of 1 to 2 metres above the ground. Signposts set up by experienced bees – that is, the scent released in the air during their flight – perhaps disperses more slowly in thick vegetation. The disadvantage of the confusing multitude of optical landmarks, so important in open country, could perhaps be compensated by their preventing the rapid dispersion of the olfactory signposts. Here, too, a fascinating area of research is opening up in forest bee biology.

Epilogue

Honey Bees: A Way of Life

Are honey bees self-aware? A conversation

DS *Are bees self-aware? Or, to put it another way, what do you think about assigning intellectual abilities and even human characteristics to bees?*

JT An essential feature of science is using clearly defined and generally accepted terms. However, such terms are often unwieldy and incorporate highly complex material that only a narrow group of specialists can appreciate. To present factual information to a broader public, terms based on everyday experience can quickly establish a particular concept. This should not stray too far from a clear and professional representation of the facts, but, as in this book, it is sometimes unavoidable and the result can be a slightly distorted view.

DS *That sounds like a cautious criticism. Can you provide a concrete example?*

JT We used the term 'selfish genes' in connection with the origin of social behaviour in honey bees. A gene itself cannot be 'selfish' in a personal sense, but the effect and the mechanism of particular gene mutations on individuals can appear as though the new gene has the characteristic of being selfish. Such a formulation aptly summarises a highly complex matter. During evolution there were, and are, advantageous genes, providing their carriers with characteristics promoting the raising of offspring. For example, humans possessing a gene enabling them to digest milk as adults spread throughout northern Europe, because when they began to keep cattle, they were able to drink the milk as well as consume the flesh and blood of their livestock. In times of hunger this was of great importance, because they had a renewable source of nourishment. Consequently, more offspring became carriers of this gene and today lactose intolerance in Europe is relatively rare in comparison to Asia. So is the gene that makes it possible to digest lactose 'selfish'? Were carriers of this gene selfish? Or were they simply better adapted to their environment?

DS *I think I understand what you are getting at. Could you explain this in relation to honey bees?*

JT Let me take behavioural biology as a way to bring us closer to bees. The goal-oriented behaviour and behavioural strategies of bees are often mentioned. However, one can safely assume that organisms such as bees cannot really follow a goal or conceive a strategy. Animals in general, though, and bees in particular, appear to us to have planned an outcome when we see that they have achieved it. In their competition with one another, those that reproduce more successfully look as if they have consciously followed a particular strategy to achieve this.

DS *That all sounds a little academic. Can bees think, plan and be aware of themselves?*

JT I don't wish to be too forthright, but I recognise your abilities to think, plan and be self-aware from the information that I receive from and about you, and transfer to you the properties that I recognise in myself. I believe this to be justified because we are very similar biological organisms.

In bees, this is a little different. When something appears 'as if' (as in the case of the selfish genes), there must be convincing evidence available that this 'as if' is not just an analogy of our own behaviour but represents the truth about the mechanisms and internal processes concerned. Bees sometimes behave 'as if' they can think and plan and are aware of themselves. We cannot say more.

In no way does this detract from the fascination of bees – quite the opposite. Nature has produced a superorganism in the form of a bee colony that does (nearly) everything right, even if it cannot think and plan and its members are not self-aware. This does not demystify bees, but in my mind deepens my awe and respect for nature and these insects.

*

It is the end of January and a high-pressure system has brought sunny winter weather. The temperature at night lies around −5°C. During the day, in the sun, it rises above freezing. What are the bees doing? The sunlight falls on the hive entrance and, indeed, the light and warmth lures the first bees out. Some even take off and take a turn around the hive. Can one already feel warmth in the hive? Remove the lid and place a hand carefully onto the foil covering the winter cluster. The bees are not packed quite so closely together as before. And is the temperature over the cluster not perhaps just a little higher than the surroundings? Is the first brood already being warmed?

In the northern hemisphere, bees start the new year after the winter solstice still hidden in their hives, raising the brood and gathering less often into their winter cluster. Beekeepers are overtaken by an almost painful expectancy. Another six to eight weeks and then it will be time for the spring inspection. Eight weeks! Beekeepers experience impatient anticipation, like children before Christmas, for spring and the start of the new bee season.

'Those who begin to keep bees and, after the dramas of the first three years still have them, no longer possess the bees, the bees possess them.' A relationship with the Bien superorganism completely overwhelms one. For most beekeepers, beekeeping is not just a hobby, it is a way of life, a way of seeing the world.

Perhaps this is because keeping bees demands and schools all our senses. We hear the calming hum of a contented colony or the aggressive whine of an angry field bee. Bit by bit, the seasonal labour of the colonies teaches us to understand the language of bees. The tone of a colony betrays immediately how they are faring, whether a bumblebee or a hornet is approaching, or even another bee. Those who have learnt to listen to bees can determine all this from the sounds they make.

And as one learns to hear more acutely, so the feel and touch of bees becomes a different experience. Naturally one is aware of stings. They are just a part of it all and one learns to be careful, to grasp the frames slowly and discover that a gentle hand can be a welcome resting place for a honey maker. Bees let themselves be stroked, if one knows how.

Colony inspections are very sticky occasions and it is irritating when you forget water for cleaning. On the other hand, honey tastes good, and is different from year to year. Honey bees expose us to a panoply of aromas and tastes for our noses and tongues. The smell of propolis; the taste of dandelion, chestnut or honeydew. Beekeepers learn to savour the different seasons.

And they learn to see them in a different light. There is a popular internet game based on identifying various cars from pictures of their tail lights and it's surprising how often we get them correct. Studies have shown that people are able to recognise many car types from a picture of a very small portion of the vehicle. But perhaps this is not so surprising. We notice what we are used to seeing. Those whose world consists of cars, offices and living rooms recognise all that is seen from these perspectives. Tail lights, instead of trees. How many of us can distinguish five different kinds of trees from their bark, fruit or leaves?

Novice beekeepers quickly learn much that is new and gain a completely different view of the natural world that surrounds them. Suddenly everything that blooms or grows is interesting because it is important to know if bees can find food at a particular location or not. All at once, one is preoccupied with the seasonal sequence of blossoming flowers, bushes and trees. Will there be an interruption to the honey flow? Will the bees have to be moved or even fed?

Over time, most beekeepers develop a lively understanding of and an awakened consciousness about the passage of the seasons. No two years are the same; no summer is like another. The question is not how often the sun shines or if it rains. Wet summers can mean vital colonies and an excellent honey harvest. The fascination comes from watching, puzzling and wondering.

Bees teach us to be amazed by what surrounds us. The forest goblin-like characteristics of some beekeepers may be due to their being adults that continue to experience life the way they did as children. Bees reveal to them a world that is filled with surprises, secrets and riddles. Bees bring them, in the dull realities of our time, a feeling of the mystery of existence and that everything is somehow bound up with everything else.

Beekeeping has a spiritual dimension. Bees lead one to the big questions. Beekeeping means watching over the lives of others and realising that their lives are not a domesticated opposite of our own, but instead held in mutual trust. Bees teach us what Albert Schweitzer expressed: that we are united with all life and that there is a universal will to live.

And does the special form of communal life that bees have developed during their evolution perhaps have a message for us? Do bees convey an idea of how to live with one another? Bees huddle together in winter clusters on combs containing honey or food provided by beekeepers. Here, they move not only around one another but also through the food stores. The cluster migrates on from food stores that are emptied to those that are still full, and is always in contact with food. Bees in contact with capped honey cells during their circulation through the cluster gnaw the caps open (the wax falls to the hive floor for the beekeeper to clean up in spring), take up honey and share it out, for not every bee in the cluster comes into direct contact with the comb. No-one goes hungry. Those that have, share – always and as long as there is something to share. A bee colony therefore starves over a few hours, not over weeks. There are no poor or weak who go without when supplies are low, or any strong, rich and without scruple who take what they need from others. Bees do not place their capital in the hands of the few. They share what they have, and when there is no more they all die together, which can be a great loss for beekeepers. Apiarists that have not noticed on a day at noon that a colony is without food, and do not immediately feed them, will open the hive on the next day and find no single bee alive.

The winter cluster is a small example of the general principle that governs the bee community – an unconditional and mutual sharing. Only while all care for one another in the 'knowledge' that they will be cared for themselves, can they be a superorganism. Is this the message from bees? To hold up a world to us in which the Golden Rule – the principle of treating others the way we would want to be treated – is a lived reality?

Many who are closely associated with bees have such thoughts. And even if this beekeeper pathos is a little overdrawn here, getting to know a small world in which cooperation functions helps to counter the dismay over a world that appears to be disintegrating through ever-increasing egotism. Bees give us hope – and bring us joy.

Acknowledgements

This book is not only our work. We owe thanks to many who shared their knowledge with us, encouraged and supported us. We extend our thanks to all those in the Gütersloher publishing house. Ralf Markmeier, as publisher, who made it possible to publish a book by a colleague. Nicole Neumann, with boundless patience was responsible for impeccable typesetting. Sigrid Fortkord accompanied and encouraged us as our reader. Gudrun Krieger oversaw the corrections with professional care and all the Gütesloher colleagues thoroughly engaged themselves with this project. Their enthusiasm carried us through!

I, Diedrich Steen, thank those who taught me beekeeping. My father, Dirk Steen, who inherited three bee colonies from my grandfather, Abram Strohschneider, and so brought bees into the

ACKNOWLEDGEMENTS

home, and Johann Noichl, who guided me to the right path after my catastrophic start. And Dr Gerd Liebig, who with his book *Einfach Imkern* and several meetings firmly established the way. I thank Pia Aumeier, whose courses I continue to experience as a steadily flowing source of insights. And I thank Michael Schlangenotto, Bruno Gründtkemeier, Franz Austermann, Siegfried Timm, Reinhard Diekhans and Dr Alexander Lojewski. My continuous exchanges with all of you prevented me from much beekeeping stupidity. A special thanks to my wife, Inge Bohnke-Steen. She not only, more or less, gave me my first bees and, more or less, endured the fact that on separator days 'everything was sticky'. She also encouraged me to undertake this book and put up with the reduced time we had to spend together due not only to the demands of my career, but also because of the book.

I, Jürgen Tautz, thank all those that awakened my very late interest in honey bees, which unfolded into a fascinating area of research. Martin Lindauer for the crafty gift of a bee colony and the reproach, said with a twinkle in his eye, that it is an error for zoologists not to interest themselves in bees. I thank my wife, Rosemarie Müller-Tautz, and my children, Meiko, Silke and Mona, for their, to my surprise, endless understanding for my work and the not always easily endured effect on the immediate surroundings. I thank all my colleagues, students, beekeepers and other bee friends who feel included, for their encouragement and contributions.

Bibliography

F.G. Barth, *Biologie einer Begegnung: Die Partnerschaft der Insekten und Blumen*, Deutsche Verlags-Anstalt Stuttgart, 1982.

R. Basile, C.W. Pirk & J. Tautz, 'Trophallactic activities in the honey bee brood nest – Heaters get supplied with high performance fuel', *Zoology*, vol. 111, 2008, pp. 433–41.

D. Bauer & K. Bienenfeld, 'Hexagonal combs of honey bees are not produced by a liquid equilibrium process', *Naturwissenschaften*, vol. 100, 2013, pp. 45–9.

M. Beye & M. Hasselmann, 'Männchen, die keiner Vater haben. Neues von der geschlecterentwicklung am Beispiel der Biene', *Biologen heute*, vol. 2, 2004, pp. 3–9.

F. Bock, *Untersuchungen zu natürlicher und manipulierter Aufzucht von Apis mellifera: Morphologie, Kognition und Verhalten*, Dissertation, Universität Würzburg, 2005.

BIBLIOGRAPHY

B. Bujok, *Thermoregulation im Brutbereich der Honigbiene Apis mellifera carnica*, Dissertation, Universität Würzburg, 2005.

L. Chittka & J. Tautz, 'The spectral input to the honey bee visual odometry', *Journal of Experimental Biology*, vol. 206, 2002, pp. 2393–7.

D. Clarke, H. Whitney, G. Sutton & D. Robert, 'Detection and learning of floral electric fields by bumblebees', *Science*, vol. 340, 2013.

R. Dawkins, *The Selfish Gene*, Springer, New York, 2014 (originally published 1976).

R.J. De Marco, J.M. Gurevitz & R. Menzel, 'Variablity in the encoding of spatial information by dancing bees', *Journal of Experimental Biology*, vol. 211, 2008, pp. 1635–44.

Deutscher Imkerbund E.V., *Jahresbericht 2015/2016*, Wachtberg, 2016.

A.G. Dyer & S.K. William, 'Mechano-optical lens array to simulate insect vision photographically', *The Imaging Science Journal*, vol. 53, 2005, pp. 209–13.

H.G. Esch, 'Über die Schallerzeugung beim Werbetanz der Honigbiene', *Zeitschrift für Vergleichende Physiologie*, vol. 45, 1961, pp. 1–11.

H.G. Esch & J.E. Burns, 'Honey bees use optical flow to measure the distance to a food source', *Naturwissenschaften*, vol. 82, 1995, pp. 28–50.

H.G. Esch, S. Zhang, M.V. Srinivasan & J. Tautz, 'Honey bee dances communicate distances measured by optic flow', *Nature*, vol. 411, 2001, pp. 581–3.

M. Fehler, M. Kleinhenz, F. Klügl, F. Puppe & J. Tautz, 'Caps and gaps: A computer model for studies on brood incubation strategies in honey bees (*Apis mellifera carnica*)', *Naturwissenschaften*, vol. 94, 2007, pp. 675–80.

T.E. Ferrari & J. Tautz, 'Severe honey bee (*Apis mellifera*) losses correlate with geomagnetic disturbances in Earth's atmosphere', *Journal of Astrobiology*, vol. 134, 2015.

J.K. Fischer, T. Müller, A-K. Spatz, U. Greggers, B. Grünewald & R. Menzel, 'Neonicotinoids interfere with specific components of navigation in honey bees', *PLOS ONE*, 2014.

G. Friedmann, *Bienengemäß imkern – das Praxishandbuch*, BLV Buchverlag, München, 2016.

K. von Frisch, *Tanzsprache und Orientierung der Bienen*, Springer Verlag, Berlin, Heidelberg, New York, 1965.

K. von Frisch & R. Jander, 'Über den Schwänzeltanz der Bienen', *Zeitschrift fur Vergleichende Physiologie*, vol. 40, 1957, pp. 239–63.

K. von Frisch & M. Lindauer, *Aus dem Leben der Bienen*, Springer Berlin, Heidelberg, New York, 1993.

B. Fröhlich, J. Tautz & M. Riederer, 'Chemometric classification of comb and cuticular waxes of the honey bee *Apis mellifera carnica*', *Journal of Chemical Ecology*, vol. 26, 2000, pp. 123–37.

H. Gätschenberger, T. Azzami, J. Tautz & H. Beier, 'Antibacterial immune competence of honey bees (*Apis mellifera*) is adapted to different life stages and environmental risks', *PLOS ONE*, vol. 8, 2013, e66415.

J.L. Gould, 'Honey bee communication: The dance language controversy', PhD thesis, Rockefeller University, New York, 1975.

U. Greggers, G. Koch, V. Schmidt, A. Dürr, A. Floriou-Servou, D. Piepenbrock, M.C. Göpfert & R. Menzel, 'Reception and learning of electrical fields in bees', *Proceedings of the Royal Society B*, 2013.

H.J. Gross, M. Pahl, A. Si, H. Zhu, J. Tautz & S. Zhang, 'Number-based visual generalisation in the honey bee', *PLOS ONE*, vol. 4, no. 1, 2009.

Gymnasium Wendelstein, Schwänzeltanz und Brauseflugkommunikation der Bienen im Stock und an der Futterstelle. Ein Beitrag zur Überprüfung einer Darstellung auch in Schulbüchern, Wettbewerbsbeitrag für 'Jugend forscht', 2017.

W.D. Hamilton, *Narrow Roads of Gene Land, Volume 1: The Evolution of Social Behaviour*, Oxford University Press, 1996.

B. Heinrich, *The Hot-Blooded Insects: Strategies and Mechanisms of Thermoregulation*, Springer, Berlin, Heidelberg, New York, 1993.

R. Hepburn, *Honey Bees and Wax*, Springer, Berlin, Heidelberg, New York, 1986.

E. Herold & K. Weiß, *Neue Imkerschule. Theoretisches und praktisches Grundwissen*, 9. Auflage, München, 1995.

K. Hoppenhaus, 'Die Biene und das Biest', *Die Zeit*, no. 44, 2011, 49f.

W. Kaiser, 'Busy bees need rest, too: Behavioural and electromyographical sleep signs in honey bees', *Journal of Comparative Physiology A*, vol. 163, 1988, pp. 565–84.

B. Karihaloo, K. Zhang & J. Wang, 'Honey bee combs: How the circular cells transform into rounded hexagons', *Interface: Journal of The Royal Society*, vol. 10, 2013.

B.A. Klein & T.D. Seeley, 'Work or sleep? Honey bee foragers opportunistically nap during the day when forage is not available', *Animal Behaviour*, vol. 82, 2011, pp. 77–83.

B.A. Klein, M. Stiegler, A. Klein & J. Tautz, 'Mapping sleeping bees within their nest: Spatial and temporal analysis of worker honey bee sleep', *PLOS ONE*, 2014.

M. Kleinhenz, *Die Wärmeübertragung im Brutbereich der Honigbiene (Apis mellifera)*, Dissertation, Universität Würzburg, 2008.

U. Kreutzer, *Karl von Frisch. Eine Biographie*, München, 2010.

T. Landgraf, R. Rojas, H. Nguyen, F. Kriegl & K. Stettin, 'Analysis of the waggle dance motion of honeybees for the design of a biomimetic honey bee robot', *PLOS ONE*, vol. 6, 2011, pp. 1–10.

M. Lindauer 'Martin Lindauer describes his experiment in "Lindauer unpublished" on pages 138 and 139 in B. Höldobler & M. Lindauer, *Fortschrittte Zoologie*, vol. 31, 1985.

M. Lindauer, 'Ein Beitrag zur Frage der Arbeitsteilung im Bienenstaat', *Journal of Comparative Physiology*, vol. 34, 1952, pp. 299–345.

M. Lindauer, 'Schwarmbienen auf Wohnungsuche', *Journal of Comparative Physiology*, vol. 37, 1955, pp. 263–324.

K. Lunau, Chr. Verhoeven, 'Wie Bienen Blumen sehen – Falschfarbenaufnahmen von Blüten', *Biologie in Unserer Zeit*, vol. 2, 2017.

M. Maeterlinck, *Das Leben der Bienen*, Eugen Diederichs Jena, 1919.

H. Martin, 'Zur Nahorientierung der Biene im Duftfeld, zugleich ein Nachweis für die Osmotropotaxis bei Insekten', *Journal of Comparative Physiology*, vol. 48, 1964, pp. 481–533.

H. Martin & M. Lindauer, 'Sinnesphysiologisches Leistungen beim Wabenbau der Honigbiene', *Journal of Comparative Physiology*, vol. 53, 1966, pp. 372–404.

J. Melcher, M. Kramer, J. Heinrich, J. Günster & J. Tautz, 'Verfahren zur Herstellung keramischer Konstruktionselemente in Nano-bis Zentimeter-Bereich', German Patent DA 10 2005 025367.9, 2005.

R. Menzel & M. Eckoldt, *Die Intelligenz der Bienen. Wie sie denken, planen, fuhlen und was wir daraus lernen können*, Knaus, München, 2016.

R. Menzel, R. Kirbach, W.-D. Haas, B. Fischer, J. Fuchs, M. Koblofsky, K. Lehmann, L. Reiter, H. Meyer, H. Nguyen, S. Jones, P. Norton & U. Greggers, 'A common frame of reference for learned and communicated vectors in honey bee navigation', *Current Biology*, vol. 21, 2011, pp. 645–50.

R.F.A. Moritz & E.E. Southwick, 'Bees as superorganisms. An evolutionary reality', Springer, Berlin, Heidelberg, New York, 1992.

K. Münstedt, S. Hofmann & K.P. Münstedt, 'Bienenprodukte in der Medizin. Apitherapie nach wissenschaftliche Kriterien bewertet', Aachen, 2015.

P. Neumann & T. Blacquiere, 'The Darwin cure for apiculture? Natural selection and managed honey bee health', *Evolutionary Applications*, vols 1–5, 2016.

J. Nitschmann & O.J. Hüsing, 'Lexicon der Bienenkunde', *Tosa Wien*, 2002.

C. Novottnick, 'Die Honigbiene. Die neue Brehm Bücherei', *Westarp Wissenschaften*, Magdeburg, 2004.

M. Pahl, J. Tautz & S. Zhang, 'Honey bee cognition' in P. Kappeler (ed.), *Behaviour: Evolution and mechanisms*, Springer, Berlin, Heidelberg, New York, 2007.

M. Pahl, H. Zhu, W. Pix, J. Tautz & S.W. Zhang, 'Circadian-timed episodic-like memory – a bee knows what to do when, and also where', *Journal of Experimental Biology*, vol. 210, 2007, pp. 3559–67.

J. Pieper, *Über die Liebe*. Kösel München, 2014

C.W.W. Pirk, H.R. Hepburn, S.E. Radloff & J. Tautz, 'Honey bee combs: Construction through liquid equilibrium process?', *Naturwissenschaften*, vol. 91, 2004, p. 353.

C.W.W. Pirk, P. Neumann, H.R. Hepburn, R.A. Moritz & J. Tautz, 'Egg viability and worker policing in honey bees', *Proceedings of the National Academy of Science*, vol. 101, 2004, pp. 8649–51.

A.M. Reynolds, J.L. Swain, A.D. Smith, A.P. Martin & J.L. Osborne, 'Honey bees use a Levy flight search strategy and odour-mediated anemotaxis to relocate food sources', *Behavioural Ecology and Sociobiology*, vol. 64, 2009, pp. 115–23.

J.R. Riley, U. Greggers, A.D. Smith, D.R. Reynolds & R. Menzel, 'The flight paths of honey bees recruited by the waggle dance', *Nature*, vol. 435, 2005, pp. 205–7.

K. Rohrseitz, *Biophysikalische und ethologische Aspekte der Tanzkommunikation der Honigbiene (Apis mellifera Pollm.)*, PhD thesis, Universität Würzburg, 1998.

F. Ruttner, *Naturgeschichte der Honigbienen*, Ehrenwirth, München, 1992.

D.C. Sandeman, J. Tautz & M. Lindauer, 'Transmission of vibration across honey combs and its detection by bee leg receptors', *Journal of Experimental Biology*, vol. 199, 1996, pp. 2585–94.

T. Schiffer, *Beenature Projekt zur Erforschung des Büchrskorpions in der Imkerei*, http://beenature-project.com, 2017.

BIBLIOGRAPHY

C.H. Schneider, J. Tautz, B. Grünewald & S. Fuchs, 'RFID tracking of sublethal effects of two neonicotinoid insecticides on the foraging behavior of *Apis mellifera*', *PLOS ONE*, vol. 7, no. 1, 2012.

T.D. Seeley, *Honey Bee Ecology: A study of adaptation in social life*, Princeton University Press, Princeton, 1985.

T.D. Seeley, *The wisdom of the hive: The social physiology of honey bee colonies*, Harvard University Press, Cambridge, 1995.

T.D. Seeley, *Honigbienen. Im Mikrokosmos des Bienenstocks*, Birkhäuser, Basel, Boston, Berlin, 1997.

T.D. Seeley, *Bienedemokratie – Wie Bienen kollektiv entscheiden und was wir davon lernen können*, Stuttgart, 2014.

T.D. Seeley, M. Kleinhenz, B. Bujok & J. Tautz, 'Thorough warm-up before take-off in honey bee swarms', *Naturwissenschaften*, vol. 90, 2003, pp. 256–60.

T.D. Seeley & J. Tautz, 'Worker piping in honey bee swarms and its role in preparing for liftoff', *Journal of Comparative Physiology*, vol. 187, 2001, pp. 667–76.

M. Simone-Finstrom & M. Spivak, 'Propolis und Bienengesundheit: Die Naturgeschichte und die Bedeutung des Gebrauchs von Pflanzenharzen durch Bienen', *Apidologie*, vol. 41, 2010, pp. 295–311.

M. Srinivasan, S.W. Zhang, M. Altwein & J. Tautz, 'Honey bee navigation: Nature and calibration of the "odometer"', *Science*, vol. 287, 2000, pp. 851–3.

J. Storm, *The Dynamics and Flow Field of the Wagging Dancing Honey Bee*, PhD thesis, Odense University, Denmark, 1998.

M. Strehle, F. Jenke, B. Fröhlich, J. Tautz, M. Riederer, W. Kiefer & J. Popp, 'A Raman-spectroscopic study of the spatial distribution of propolis in the comb of *Apis mellifera carnica*', *Biospectroscopy*, vol. 72, 2003, pp. 217–24.

S. Su, S. Albert, S. Zhang, S. Maier, S. Chen, H. DFu & J. Tautz, 'Non-destructive genotyping and genetic variation of fanning in a honey bee colony', *Journal of Insect Physiology*, vol. 53, 2007, pp. 411–17.

J. Tautz, 'Das Festnetz der Bienen', *Spektrum der Wissenschaft*, August 2002, pp. 60–6.

J. Tautz, *Der Bien – Superorganismus Honigbiene*, 2 Audio-CDs, 144 Minuten + Fotobooklet, (Regie: Klaus Sander). Köln, 2007.

J. Tautz, *Phänomen Honigbiene*, Spectrum Akademisches Verlag, 2007.

J. Tautz, *Die Erforschung der Bienewelt. Neue Daten – neues Wissen*. Klett MINT Verlag, 2015.

J. Tautz & M. Lindauer, 'Telefonnetz und chemische Gedächtnis: Wachs als vielseitiges Kommunikationsmediium der Honigbiene', *Akademie-Journal* (Mainz), vol. 1, 1999, p. 15.

J. Tautz & K. Rohrseitz, 'What attracts honey bees to a waggle dancer?' *Journal of Comparative Physiology A*, vol. 183, 1998, pp. 661–7.

J. Tautz, K. Rohrseitz & D.C. Sandeman, 'One-strided waggle dance in bees', *Nature*, vol. 382, 1996, p. 32.

J. Tautz & M. Rostas, 'Honey bee buzz attenuates plant damage by caterpillars', *Current Biology*, vol. 18, 2008, R1125–R1126.

J. Tautz & D.C. Sandeman, 'Recruitment of honey bees to non-scented food sources', *Journal of Comparative Physiology A*, vol. 189, no. 4, 2002, pp. 293–300.

J. Tautz, S. Zhang, J. Spaethe, A. Brockmann, A. Si & M. Srinivasan, 'Honey bee odometry: Performance in varying natural terrain', *PLOS Biology*, vol. 2, 2004, pp. 915–23.

C. Thom, T.D. Seeley & J. Tautz, 'A scientific note on the dynamics of labor devoted to nectar foraging in a honey bee colony: Number of foragers versus individual foraging activity', *Apidologie*, vol. 31, 2000, pp. 737–8.

L. Tison, M.L. Hahn, S. Holtz, A. Rößner, U. Greggers, G. Bischoff & R. Menzel, 'Honey bees' behaviour is impaired by chronic exposure to the neonicotinoid thiacloprid in the field', *Environmental Science & Technology*, 2016.

T. Tourneret & S. de Saint Pierre, *Die Wege des Honigs*, Ulmer, Stuttgart, 2017.

G. Tribe, J. Tautz, K. Sternberg & J. Cullinan, 'Firewalls in bee nests – survival value of propolis walls of wild Cape honey bee (*Apis mellifera capensis*)', *The Science of Nature*, 2017 (in press).

R. Wehner, 'Polarization vision – a uniform sensory capacity?' *Journal of Experimental Biology*, vol. 204, 2001, pp. 2589–96.

A. Weidenmüller, C. Kleineidam & J. Tautz, 'Collective control of nest climate parameters in bumblebee colonies', *Animal Behaviour*, vol. 63, 2002, pp. 1065–71.

T. Weippl, *Archiv für Bienenkunde*, vol. 9, 1928, pp. 70–9.

BIBLIOGRAPHY

A.M. Wenner & P.H. Wells, *Anatomy of a Controversy: The question of a dance 'language' among bees*, Columbia University Press, New York, 1990.

M. Winston, *The Biology of the Honey Bee*, Harvard University Press, Cambridge Mass., 1987.

W. Wu, A.M. Moreno, J.M. Tangen & J. Reinhard, 'Honey bees can discriminate between Monet and Picasso paintings', *Journal of Comparative Physiology A*, vol. 199, 2013, pp. 45–55.

Y.K. Yeshkov & A.M. Sapozhnikov, 'Mechanism of generation and perception of electric fields by honey bees', *Biofizika*, vol. 21, 1976, pp. 1097–102.

S. Zhang, S. Schwarz, M. Pahl, H. Zhu, H & J. Tautz, 'Honey bee memory: A honey bee knows what to do and when', *Journal of Experimental Biology*, vol. 209, 2006, pp. 4420–8.

Video resources

V. Klein, Thermal Society: A series of sequences (total length: 5 minutes and 55 seconds) on the use of warmth in a bee colony, 2012: www.youtube.com/watch?v=iYr158rwLBI

HOBOS team videos of the landing of experienced foragers: the buzzing flight of a dancer: www.youtube.com/watch?v=wMt74yoA4eo

Calm and quick landing of a forager that had not previously danced: www.youtube.com/watch?v=n-ACYdQ7yuA

Endnotes

Chapter 1: The Factory and Equipment of a Bee Colony

1 T. Tourneret & S. de Saint Pierre, *Die Wege des Honigs*, Ulmer, Stuttgart, 2017.

2 R. Hepburn, *Honey Bees and Wax*, Springer Berlin, Heidelberg, New York, 1986; B. Fröhlich, J. Tautz & M. Riederer, 'Chemometric classification of comb and cuticular waxes of the honey bee *Apis mellifera carnica*', *Journal of Chemical Ecology*, vol. 26, 2000, pp. 123–37.

3 M. Strehle, F. Jenke, B. Fröhlich, J. Tautz, M. Riederer, W. Kiefer & J. Popp, 'A Raman-spectroscopic study of the spatial distribution of propolis in the comb of *Apis mellifera carnica*', *Biospectroscopy*, vol. 72, 2003, pp. 217–24.

4 J. Tautz & M. Lindauer, 'Telefonnetz und chemische Gedächtnis: Wachs als vielseitiges Kommunikationsmediium der Honigbiene', *Akademie-Journal* (Mainz), vol. 1, 1999, p. 15.

5   D.C. Sandeman, J. Tautz & M. Lindauer, 'Transmission of vibration across honey combs and its detection by bee leg receptors', *Journal of Experimental Biology*, vol. 199, 1996, pp. 2585–94.

6   J. Tautz, 'Das Festnetz der Bienen', *Spektrum der Wissenschaft*, August 2002, pp. 60–6.

7   J. Tautz, K. Rohrseitz & D.C. Sandeman, 'One-strided waggle dance in bees', *Nature*, vol. 382, 1996, p. 32.

8   J. Storm, *The Dynamics and Flow Field of the Wagging Dancing Honey Bee*, PhD Thesis, Odense University, Denmark, 1998; K. Rohrseitz, *Biophysikalische und ethologische Aspekte der Tanzkommunikation der Honigbiene (Apis mellifera Pollm.)*, PhD Thesis, Universität Würzburg, 1998.

9   H. Esch, 'Über die Schallerzeugung beim Werbetanz der Honigbiene', *Zeitschrift für Vergleichende Physiologie*, vol. 45, 1961, pp. 1–11.

Chapter 2: Teamwork in the Honey Factory

1   B. Heinrich, *The Hot-Blooded Insects: Strategies and Mechanisms of Thermoregulation*, Springer, Berlin, Heidelberg, New York, 1993; B. Bujok, *Thermoregulation im Brutbereich der Honigbiene Apis mellifera carnica*, Dissertation, Universität Würzburg, 2005; M. Kleinhenz, *Die Wärmeübertragung im Brutbereich der Honigbiene (Apis mellifera)*, Dissertation, Universität Würzburg, 2008.

2   J. Tautz, *Die Erforschung der Bienenwelt. Neue Daten – neues Wissen*, Klett MINT Verlag, 2015.

3   R. Basile, C.W. Pirk & J. Tautz, 'Trophallactic activities in the honey bee brood nest – Heaters get supplied with high performance fuel', *Zoology*, vol. 111, 2008, pp. 433–41.

ENDNOTES

4   Kleinhenz, 2008.

5   H. Martin & M. Lindauer, 'Sinnesphysiologisches Leistungen beim Wabenbau der Honigbiene', *Journal of Comparative Physiology*, vol. 53, 1966, pp. 53, 372–404; D. Bauer & K. Bienenfeld, 'Hexagonal combs of honey bees are not produced by a liquid equilibrium process', *Naturwissenschaften*, vol. 100, 2013, pp. 45–9.

6   C.W.W. Pirk, H.R. Hepburn, S.E. Radloff & J. Tautz, 'Honey bee combs: Construction through liquid equilibrium process?', *Naturwissenschaften*, vol. 91, 2004, p. 353.

7   B. Karihaloo, K. Zhang & J. Wang, 'Honey bee combs: How the circular cells transform into rounded hexagons', *Interface: Journal of The Royal Society*, vol. 10, 2013.

8   J. Melcher, M. Kramer, J. Heinrich, J. Günster & J. Tautz, 'Verfahren zur Herstellung keramischer Konstruktionselemente in Nano-bis Zentimeter-Bereich', German Patent DA 10 2005 025367.9, 2005.

9   M. Lindauer, 'Ein Beitrag zur Frage der Arbeitsteilung im Bienenstaat', *Journal of Comparative Physiology*, vol. 34, 1952, pp. 299–345.

10 W. Kaiser, 'Busy bees need rest, too: Behavioural and electromyographical sleep signs in honey bees', *Journal of Comparative Physiology A*, vol. 163, 1988, pp. 565–84.

11 R. Menzel & M. Eckoldt, *Die Intelligenz der Bienen. Wie sie denken, planen, fuhlen und was wir daraus lernen können*, Knaus, München, 2016.

12 B.A. Klein, M. Stiegler, A. Klein & J. Tautz, 'Mapping sleeping bees within their nest: Spatial and temporal analysis of worker honey bee sleep', *PLOS ONE*, 2014.

13 T. Weippl, *Archiv für Bienenkunde*, vol. 9, 1928, pp. 70–9.

14 C. Thom, T.D. Seeley & J. Tautz, 'A scientific note on the dynamics of labor devoted to nectar foraging in a honey bee colony: Number of foragers versus individual foraging activity', *Apidologie*, vol. 31, 2000, pp. 737–8.

15 F. Bock, *Untersuchungen zu natürlicher und manipulierter Aufzucht von Apis mellifera: Morphologie, Kognition und Verhalten*, Dissertation, Universität Würzburg, 2005.

16 K. von Frisch, *Tanzsprache und Orientierung der Bienen*, Springer Verlag, Berlin, Heidelberg, New York, 1965, p. 236.

17 T. Landgraf, R. Rojas, H. Nguyen, F. Kriegl & K. Stettin, 'Analysis of the waggle dance motion of honey bee for the design of a biomimetic honey bee robot', *PLOS ONE*, vol. 6, 2011, pp. 1–10.

18 H.G. Esch & J.E. Burns, 'Honey bees use optical flow to measure the distance to a food source', *Naturwissenschaften*, vol. 82, 1995, pp. 28–50.

19 J. Tautz, S. Zhang, J. Spaethe, A. Brockmann, A. Si & M. Srinivasan, 'Honey bee odometry: Performance in varying natural terrain', *PLOS Biology*, vol. 2, 2004, pp. 915–23.

20 H.G. Esch, S. Zhang, M.V. Srinivasan & J. Tautz, 'Honey bee dances communicate distances measured by optic flow', *Nature*, vol. 411, 2001, pp. 581–3.

21 K. von Frisch & R. Jander, 'Über den Schwänzeltanz der Bienen', *Zeitschrift fur Vergleichende Physiologie*, vol. 40, 1957, pp. 239–63.

22 J.R. Riley, U. Greggers, A.D. Smith, D.R. Reynolds & R. Menzel, 'The flight paths of honey bees recruited by the waggle dance', *Nature*, vol. 435, 2005, pp. 205–7.

23 von Frisch, 1965, p. 160.

24 von Frisch, 1965, p. 166.

25 von Frisch, 1965, p. 161.

26 von Frisch, 1965, S86.

27 Riley et al., 2005.

28 Esch et al., 2001.

29 R. Menzel, R. Kirbach, W.-D. Haas, B. Fischer, J. Fuchs, M. Koblofsky, K. Lehmann, L. Reiter, H. Meyer, H. Nguyen, S. Jones, P. Norton & U. Greggers, 'A common frame of reference for learned and communicated vectors in honey bee navigation', *Current Biology*, vol. 21, 2011, pp. 645–50.

30 Menzel et al., 2011.

31 M. Beye & M. Hasselmann, 'Männchen, die keiner Vater haben. Neues von der geschlecterentwicklung am Beispiel der Biene', *Biologen heute*, vol. 2, 2004, pp. 3–9.

32 S. Su, S. Albert, S. Zhang, S. Maier, S. Chen, H. DFu & J. Tautz, 'Non-destructive genotyping and genetic variation of fanning in a honey bee colony', *Journal of Insect Physiology*, vol. 53, 2007, pp. 411–17.

ENDNOTES

33 R. Menzel & M. Eckoldt, *Die Intelligenz der Bienen. Wie sie denken, planen, fuhlen und was wir daraus lernen können*, Knaus, München, 2016.

34 R. Wehner, 'Polarization vision – a uniform sensory capacity?' *Journal of Experiential Biology*, vol. 204, 2001, pp. 2589–96.

35 L. Chittka & J. Tautz, 'The spectral input to the honeybee visual odometry', *Journal of Experimental Biology*, vol. 206, 2002, pp. 2393–7.

36 H. Martin, 'Zur Nahorientierung der Biene im Duftfeld, zugleich ein Nachweis für die Osmotropotaxis bei Insekten', *Journal of Comparative Physiology*, vol. 48 1964, pp. 481–533.

37 von Frisch, 1965.

38 Y.K. Yeshkov & A.M. Sapozhnikov, 'Mechanism of generation and perception of electric fields by honey bees', *Biofizika*, vol. 21, 1976, pp. 1097–102., recently confirmed by U. Greggers, G. Koch, V. Schmidt, A. Dürr, A. Floriou-Servou, D. Piepenbrock, M.C. Göpfert & R. Menzel, 'Reception and learning of electrical fields in bees', *Proceedings of the Royal Society B*, 2013.

39 D. Clarke, H. Whitney, G. Sutton & D. Robert, 'Detection and learning of floral electric fields by bumblebees', *Science*, vol. 340, 2013.

40 T.E. Ferrari & J. Tautz, 'Severe honey bee (*Apis mellifera*) losses correlate with geomagnetic disturbances in Earth's atmosphere', *Journal of Astrobiology*, vol. 134, 2015.

ENDNOTES

41 M. Pahl, J. Tautz & S. Zhang, 'Honey bee cognition' in P. Kappeler (ed.), *Behaviour: Evolution and mechanisms*, Springer, Berlin, Heidelberg, New York, 2007.

42 S. Zhang, S. Schwarz, M. Pahl, H. Zhu, H & J. Tautz, 'Honey bee memory: A honey bee knows what to do and when', *Journal of Experimental Biology*, vol. 209, 2006, pp. 4420–8.

43 Pahl, Tautz & Zhang, 2007.

44 Pahl, Tautz & Zhang, 2007.

45 H.J. Gross, M. Pahl, A. Si, H. Zhu, J. Tautz & S. Zhang, 'Number-based visual generalisation in the honey bee', *PLOS ONE*, vol. 4, no. 1, 2009.

46 W. Wu, A.M. Moreno, J.M. Tangen & J. Reinhard, 'Honey bees can discriminate between Monet and Picasso paintings', *Journal of Comparative Physiology A*, vol. 199, 2013, pp. 45–55.

47 M. Lindauer 'Martin Lindauer describes his experiment in "Lindauer unpublished" on pages 138 and 139 in B. Höldobler & M. Lindauer, *Fortschrittte Zoologie*, vol. 31, 1985.

Chapter 3: The Honey Factory Production Line

1 Interested readers are referred to K. Münstedt, S. Hofmann & K.P. Münstedt, 'Bienenprodukte in der Medizin. Apitherapie nach wissenschaftliche Kriterien bewertet', Aachen, 2015.

2 T. Schiffer, personal communication, 2017.

3 G. Tribe, J. Tautz, K. Sternberg & J. Cullinan, 'Firewalls in bee nests – survival value of propolis walls of wild Cape honey bee (*Apis mellifera capensis*)', *The Science of Nature*, 2017 (in press).

4   Münstedt, Hofmann & Münstedt, 2015, p. 117.

5   See comments in Chapter 2, and also A.M. Wenner & P.H. Wells, *Anatomy of a Controversy: The question of a dance 'language' among bees*, Columbia University Press, New York, 1990.

6   A. Weidenmüller, C. Kleineidam & J. Tautz, 'Collective control of nest climate parameters in bumblebee colonies', *Animal Behaviour*, vol. 63, 2002, pp. 1065–71.

Chapter 4: Founding a Daughter Company

1   M. Lindauer, 'Schwarmbienen auf Wohnungsuche', *Journal of Comparative Physiology*, vol. 37, 1955, pp. 263–324; T.D. Seeley, *Bienedemokratie – Wie Bienen kollektiv entscheiden und was wir davon lernen können*, Stuttgart, 2014.

2   T.D. Seeley & J. Tautz, 'Worker piping in honey bee swarms and its role in preparing for liftoff', *Journal of Comparative Physiology*, vol. 187, 2001, pp. 667–76.

3   T.D. Seeley, M. Kleinhenz, B. Bujok & J. Tautz, 'Thorough warm-up before take-off in honey bee swarms', *Naturwissenschaften*, Vol. 90, 2003, pp. 256–60.

4   R. Dawkins, *The Selfish Gene*, Springer, New York, 2014 (originally published 1976).

5   W.D. Hamilton, *Narrow Roads of Gene Land, Volume 1: The Evolution of Social Behaviour*, Oxford University Press, 1996.

6   C.W.W. Pirk, P. Neumann, H.R. Hepburn, R.A. Moritz & J. Tautz, 'Egg viability and worker policing in honeybees', *Proceedings of the National Academy of Sciences*, vol. 101, 2004, pp. 8649–51.

ENDNOTES

Chapter 5: Bees as Aggressors

1. H. Gätschenberger, T. Azzami, J. Tautz & H. Beier, 'Antibacterial immune competence of honey bees (*Apis mellifera*) is adapted to different life stages and environmental risks', PLOS ONE, vol. 8, 2013, e66415.

Chapter 6: The Struggle for Survival

1. Gätschenberger et. al., 2013.

2. M. Simone-Finstrom & M. Spivak, 'Propolis und Bienengesundheit: Die Naturgeschichte und die Bedeutung des Gebrauchs von Pflanzenharzen durch Bienen', *Apidologie*, vol. 41, 2010, pp. 295–311.

3. Gätschenberger et. al., 2013.

4. C.H. Schneider, J. Tautz, B. Grünewald & S. Fuchs, 'RFID tracking of sublethal effects of two neonicotinoid insecticides on the foraging behavior of *Apis mellifera*', PLOS ONE, vol. 7, no. 1, 2012; J.K. Fischer, T. Müller, A-K. Spatz, U. Greggers, B. Grünewald & R. Menzel, 'Neonicotinoids interfere with specific components of navigation in honey bees', PLOS ONE, 2014; L. Tison, M.L. Hahn, S. Holtz, A. Rößner, U. Greggers, G. Bischoff & R. Menzel, 'Honey bees' behaviour is impaired by chronic exposure to the neonicotinoid thiacloprid in the field', *Environmental Science & Technology*, 2016.

5. J. Tautz & M. Rostas, 'Honey bee buzz attenuates plant damage by caterpillars', *Current Biology*, vol. 18, 2008, R1125–R1126.

6. T. Schiffer, *Beenature Projekt zur Erforschung des Büchrskorpions in der Imkerei*, http://beenature-project.com, 2017.

Index

Achard, Franz Carl 11
afterswarms 170–2
alarm pheromone 58
Albert Schweitzer 235
allergy 122
anaphylactic shock 122
antennae 24–5, 46–7, 62, 64, 110, 141, 147
    and heater bees 45
    and the queen 86
    and scent, odour, perfume 45
    and swarming 165
aphids 134–5, 153
apiarists 5, 94, 99, 107, 126, 181, 193, 196, 207, 236
    see also basket beekeepers; beekeepers; hobby apiarists
    and genus *Apis* 220
    and hand-feeding bees 153
    and honey collection 143–4, 148, 154
    and magazine hives 218
    and the Varroa mite 201
Apis cerana 187–8
Apis genus 220
Apis mellifera capensis 129, 187–8, 202
apitherapy 119–20, 129, 146
apitoxin 121, 123–4 *see also* bee venom

INDEX

ballast 154
baseboard 14, 16, 22, 169 *see also* magazine hives
basket beekeepers 10–13, 14, 172
basket hives 9–11, 13, 169 *see also* skeps
bee blood *see* haemolymph
bee bread 40, 120, 132
bee caves 9, 17
bee colony 3, 8, 33, 36, 94–6, 186, 213, 217 *see also* learning capacity of bees
bee colony superorganism 104–5, 116, 198, 210, 224, 232–3, 236 *see also* Bien
bee comb *see* comb cells; combs
bee dance 23–7, 61, 69–70, 85, 136–8, 141, 163 *see also* comb telephone; communication; dancers; waggle dance
   and 'dance communication' experiments 67–9, 75–7
   and the swarm cluster 163–4
bee die-off/mortality 198, 200, 208, 210, 236 *see also* colony collapse disorder (CCD)
   and cold temperature 217, 221
   and honey bees 203
   and insecticide 209
   and sublethal effects 214

bee recruits 80–3 *see also* waggle dance
   and the bee dance 69–70, 138, 140–1
   and pheromones 80–1
   and research on 225–6
   and the swarm 165
bee refuge 219, 224
bee saliva 132, 144–5
bee sleep 43, 62, 62–4
bee venom 120–2, 124 *see also* apitoxin
bee vision 73, 106–9
beekeepers 5, 58, 60, 89, 120, 173–4 *see also* apiarists; basket beekeepers; city apiarists; hobby apiarists; pollination apiary/beekeeping
   and bee stings 122–3
   and brood combs 21, 27, 39, 42
   and collecting beeswax 126
   and diversity of bee species 93
   and flower constancy 142
   and harvesting bee bread 132
   and harvesting honey 148–9, 153
   and harvesting pollen 131
   and harvesting propolis 130
   and magazine hives 15–16
   and the queen bee 102–3

INDEX

and the 'queen cell test' 177
and queen replacement cells 176
and the queenless colony 99–100
and spring blossoming 52
and the Varroa mite 188–9, 195–6
beekeeping early practices 9–14, 197
beekeeping practices 202, 220–2 *see also* magazine hives; multiple storey hives; Zeidler method
beeswax 10–12, 20, 65, 124–6, 151, 193 *see also* 'cap wax'; cell-cap wax; wax plates
   and apitherapy 120
   and comb building 52, 54–5
   and thermal radiation 50–1
   and wax molecules 18
beeswax candles 10–11, 124–6
beet sugar 11
beetroot sugar syrup 133
Berlepsch, Baron August von 14–15
Bien 104, 233 *see also* bee colony superorganism
blossom phases 131, 142, 184–5, 205, 207, 209
   and spring 159
   and summer 153, 184

book scorpion (pseudoscorpion) 222–3
breeding frames 179–80
brood areas 138, 151 *see also* brood nest temperature; heater bees
   and trophallaxis 43
   and warmth 41–2, 44–51
brood cells 13, 21, 42, 47–9, 51, 100, 126, 154 *see also* heater bees; warmth
   and cleaner bees 38–9
   and formic acid treatment 194
   and robber bees 186
   and the swarm drive 174–5, 206
   and the Varroa mite 190–2
   and warmth 43–6
   and worker bees 21
brood combs 50, 111, 116, 126, 176, 212
   and beekeepers 42, 176, 178–80
   and drone mothers 177
   and honey-maker bees 56
   and humidity 146
   and the queen 39–40, 160, 168
   and queen cells 174
   and worker bees 21, 38, 101
brood nest 38, 42, 95, 100 *see also* heater bees
   and the donor colony 178

INDEX

brood nest *cont.*
    and swarming 175
    and egg accommodation 14
    and filling station bees 44
    and frames 174
    and honey harvesting 148, 150
    and honey-maker bees 144
    and nectar 41
    and nurse bees 39
    and pupal cells 49, 116
    and the queen, 103, 191
    and summer bees 47–51
    and the Varroa mite 191
    and winter bees 30
    and winter feeding 153
brood nest temperature 41, 43–4, 46–9, 51, 95, 97 *see also* heater bees; warmth
Buckfast bees 93–4
buckled brood combs 177, 186
buckwheat honey 133
builder bees 52, 54–6 *see also* comb building
bumblebees 32, 136, 147, 203, 233
buzzing flights 78–81, 86, 164, 226

cane sugar 11
canola 142–3
canola honey 133, 142
'cap wax' *see* cell-cap wax

Cape honey bee *see Apis mellifera capensis*
carbohydrates 31, 37
Carnica bees 94
Carniolian honey bee 93–4
cell caps 17–21, 30, 42, 45, 52, 154, 180, 236 *see also* comb cell edges
    and communication 25, 27
    and honey storage 145
    and old combs 126
    and removal 150
cell-cap wax 17–21, 30, 52, 126–8, 145, 150, 154 *see also* wax harvesting
central partitions 54
cherry trees 1, 41
chestnut tree honey 142
chestnut trees 128, 234
Chevreul, Michele Eugene 11
chlorophyll 108
chromosome sets 87, 91–2, 94–5
city apiarists 204–5
cleaner bees 39, 61, 64, 87, 190
clothianidin 209, 214
cocoons 38–9
cold temperature and the bee colony 8, 33, 36, 186, 213, 217, 221
colony collapse disorder (CCD) 207 *see also* bee die-off/mortality

INDEX

comb building 17–18
    and builder bees 52, 55–6
    and horizontal guides 14
    and the new bee colony 168
    and propolis 19–20
    and wooden rods 12–13
    comb cell edges 20, 22, 23, 26, 160 *see also* cell caps; comb cells; comb wax; wax plates
comb cells 21, 42, 86 *see also* brood combs; comb building; comb cell edges; comb wax; combs
    and bee food reserve 13
    and builder bees 54–5
    and communication 26–7
    and egg laying 20
    and frame size 21
    and honey harvest 150–1
    and pollen storage 132
    and propolis 19, 22
comb central partitions 21, 54, 169
comb centre 13
comb network structure 18–20, 54–5 *see also* beeswax; comb building; comb cells; comb cell edges
comb passages 32–3, 141, 147
comb telephone 22–7, 140 *see also* bee dance; communication; waggle dance

comb vibration 22–3, 137, 140–1, 165 *see also* communication
comb wax 17–21, 26, 50–2, 54–5, 126, 160, 162, 168, 193 *see also* beeswax; comb cells; propolis; wax harvesting; wax plates
combs 17, 19, 50 *see also* comb cells; comb vibration; comb wax; wax plates
    and bee colony survival 9
    and the nursery 20
    and communication network 22–4
    and removable frames 15–16
    and heat-sensitive cameras 54
    and honey collection 14
    and removable frames 12
    and role in egg laying 13
    and the spring breakthrough 38
    and winter 31
communication 22, 26, 85, 136–8, 141, 163 *see also* comb telephone; waggle dance
    and the bee dance 23–7, 61
    and 'dance communication' experiments 68–73, 76
    and the queen 22, 170
    and scent, odour, perfume 22
    and the swarm 6, 166

INDEX

court bees 39, 90, 99, 102, 104, 161
crystallisation 152
cultivation 181, 205, 218–19

dance
    angle 71, 136–7
    figure 69, 76, 136–7
    frequency 23–4, 26, 75, 137
    information 23, 61, 69–70, 85, 138, 141, 163
    language 69
    model 68, 76–7, 81, 85–6
    movements 23–4, 68–9, 71, 73–5, 140
    path 71–4, 136–7, 141
    pattern 71, 74
    success 79
    vitality 23–6, 71–5, 80
'dance communication' experiments 68–79
dancers 24–6, 69, 72, 74–7, 80, 82, 136–8, 140–1 *see also* bee dance; waggle dance
dandelion 142, 234
dandelion blossom 131
decoy combs 196
Deutsche Imkerbundes (DIB) 199
dextrose 145, 153
directional fan 70, 76 *see also* fanning

discarded cap wax 126, 226
drone assembly areas 94, 102–3, 107, 110 *see also* drone congregation
drone brood cells 21, 26, 90, 186, 195–6
drone combs 21, 87
drone congregation 88, 92–3, 107
drone elimination 90, 178
drone larvae 21, 92
drone mothers 93, 177
drones 21, 86–8, 93–4, 90, 98, 100, 102–4, 107, 110, 178
    see also drone assembly areas; drone elimination
    and buckled brood combs 186
    and chromosome sets 91–2
    and early life 86–8
    and the queen 89
    and resistance to disease 213
    and the Varroa mite 189–90, 195
    and worker bees 117, 177–8
Dzierzon, Johann 14–15, 87

earth's magnetic field 111
egg laying 2 *see also* larvae; nurse bees
    and comb cells 13, 20, 39–41, 49

INDEX

and drone cells 86–7, 90
and the honey area 149
and the queen 95, 103, 153, 158, 171, 175
and spring breakthrough 36–7
and winter bees 29–30
and worker bees 177–9
energy (food) reserves 8, 13, 31, 33, 37, 162, 224 *see also* winter reserves
enzymes 18, 132, 144–5
epigenetics 94–6
episodic memory 113–14
European dark bee 93

fanning 161 *see also* directional fan; ventilator bees
and brood nest temperature 95–6, 148
and 'dance communication' experiments 70, 76
feeder super 153 *see also* super
feeders 85–6, 116–17, 215, 226 *see also* buzzing flights
and bee recruits 80, 82–4, 167
and bee vision 107
and buzzing flights 78–9
and 'dance communication' experiments 69–73, 75–7
feelers 54 *see also* antennae

fermentation 132, 151
'fertilisation mark' 89
field bees 57–62, 136, 138, 169, 171, 173
and bee recruits, 80
and beekeepers 149, 153, 181
and resistance to disease 213
filling station bees 44
flight muscles 23, 165
flower blossom honey 152
flower constancy 139, 142
flower scent/perfume 80–1, 85, 109–10, 112, 136, 184
flying in 58–9, 111, 143, 153
followers 23–5, 68, 70, 140–1, 226–7
and dancers 80, 82, 138
food combs 13, 90, 154
foragers 59, 62, 64, 110, 163
and 'dance communication' experiments 67–71, 76–80
and foraging flights 30, 65–6
and gathering food 1, 31, 36–7, 131, 173
and comb telephone 140
and learning capacity of bees 116–17
and mandibular glands 128
and nectar 56–7, 138–9, 144–5
and research on 82–5, 226

269

INDEX

foragers *cont.*
    and robber bees 185–6
    and scent trails 60–1
    and swarming 168
forest 144, 172, 220–1, 223, 227
forest honey 133, 135
formic acid 193–6, 207
foulbrood 210
frame size 15–16, 21
frames 16–17, 21, 52, 126, 148
    see also breeding frames; super
    and beekeeping practices 233
    and central partitions 54, 169
    and communication 25, 27
    and hive boxes 180
    and honeycomb 150–1
    and removable frames 12, 15, 148
    and swarm prevention 173–4
Frisch, Karl von 68–81
fructose 145, 152–3
fumigation 193

geraniol 78
German normal size 16
greenhouse effect 50–1
guard bees 1, 57, 185–6
    and the alarm pheromone 58
    and being bribed 59, 224
    and drones 90

haemolymph 187, 189, 191
hand-feeding bees 153
heater bees *see also* brood nest temperature
    and brood cells 42–5
    and brood combs 95
    and the brood nest 47–2
    and comb building 55
heather 142–3, 172
heather honey 10
heat-sensitive camera 43–4, 54, 166
herbicide 212
hive air therapy *see* apitherapy
hive air/atmosphere 146–9, 157, 161
hive bees 31, 52, 64
hive blocked potential 173, 175
hive boxes 58, 148, 180
hive cases 153, 194
hive entrance 13, 22, 58, 78, 113, 139, 150, 163, 192, 224
    and drones 90
    and guard bees 57, 123
    and pollen 131–2
    and honey-maker bees 56
    and humidity 146
    and midday rest 62
    and play flights 59, 102
    and the queen 157
    and travelling bees 143

INDEX

and propolis 127, 129
and ventilator bees 95
and the waggle dance 137
hive harmony 99, 110
hive tool 128
hobby apiarists, 93, 200–3
honey 9, 39, 120, 133–5, 146, 152, 154 *see also* nectar
honey area 148–50, 153
honey collection 8, 14–15, 149–51 *see also* honey harvest
honey crescent 13–14
honey floor 56
honey flow 41, 155, 157, 185
    and comb building 52
    and honey harvest 148
    and travelling bees 143
honey flow beginning 41, 149
honey flow interruption 234
honey flow pursuit 143
honey flow seasons 155
honey flow sources 173
honey harvest 148–9, 235 *see also* honey collection
    and beekeeping early practices, 9–10
    and the separator 150
honey ripening 145–7
honey stomachs 59, 61, 118, 136, 144, 162, 171, 224–5

honey super, 14–15, 149–51 *see also* super
honey varieties 142 *see also* nectar sources
honeycomb 148, 150
honeydew 135, 142, 153–4, 234
honeydew honey 133, 135, 152
honey-maker bees 233 *see also* summer bees
    and bee sleep 64
    and field bees 59
    and foragers 56–7, 61
    and honey 149–50
    and nectar 144
humidity 128, 145–7, 148

Imhoof, Markus 205
immune system 121–2, 132, 208, 210–11
imprinted tradition 104, 116
inhibin 145
initial swarm 175 *see also* swarm drive
insecticide 209, 212
intermittent heating 34

Kaiser, Walter 62
Kepler, Johannes 53, 55

INDEX

landing boards 1, 58, 96, 157
larvae 2, 21, 38, 100, 177–8, 213
   and beekeepers 179–80
   and drones 92
   and epigenetics 95
   and nourishment 31, 41
   and nurse bees 37, 158–9, 173, 175
   and nurse's milk 39, 176
   and pollen 132
   and royal jelly 101, 124, 160
   and the Varroa mite 190–1
   and worker bees 117
larval cell 189
larval development 92, 95
larval diet 37, 39, 101, 124, 132, 160, 176 *see also* nurse's milk; pollen; royal jelly
late summer 2, 9–10, 29, 153, 172, 194, 207–8
learning capacity of bees 105, 112–18
Lindauer, Martin 62, 116–18
linden tree honey 152
linden trees 139, 142, 144
Linnaeus, Carl 112
Louis XIV 97

magazine hive super *see* super
magazine hives *see also* baseboard; frames; super
   and beekeeping practices 16–17, 201–2, 216–18
   and the book scorpion (pseudoscorpion) 223
   and the hive entrance 127
   and swarming 172
mandibular glands 40, 99, 124, 128
mating sign 89, 102
metamorphosis 37, 191
micro-behaviour 206
midday rest 62
milk glands 39
mite pressure 207
monoculture 208
mystery of existence 235

Nasanov scent glands 78, 226
natural beekeeping 216, 219
natural comb building 21, 54
nectar 60, 139
   and drones 88
   and field bees 59
   and honey 134, 145–6
   and honey stomachs 118, 136
   and honey-maker bees 56–7, 144, 149
   and summer bees 56

INDEX

nectar gatherers 2, 31, 60–1, 66–7, 69, 138
nectar offering 10, 41, 109, 204–5
nectar sources 1, 31, 60, 76, 85, 107, 110, 112–13, 131, 135, 138, 142, 163
nectar supply 10, 16, 41, 60, 149, 173, 208
nectar-less period 184–5, 217–18
nurse bees 61, 64 see also nurse's milk
  and the brood 56
  and larvae 2, 20–1, 37, 158–60, 173, 175
  and the queen 39, 98, 100–1
  and resistance to disease 213
  and royal jelly 40, 124, 176–7
nursery 20, 22, 41
nurse's milk 39–40, 124, 176, 178, 180

old combs 126, 211, 221–2
olfaction 85–6, 109–10
overwinter 154, 194, 206

paraffin wax 11
parasitic tracheal mite 223
perga see beebread
pesticides 193, 209, 212

pheromone, scent 81, 110, 161, 164
  and geraniol 78
  and the queen 14, 99, 102–3, 149, 176–7
piping 165–6
piping scouts 166
plant protection chemicals 212–14
plasticity 168
play flight swarm 59, 102
pollen 65, 109
  and bee saliva 132
  and city apiarists 203
  and comb cells 13
  and drones 88
  and foragers 173
  and harvesting of 130–1
  and nurse bees 37, 39, 41
  and pollen paste 120
  and pollination apiary/beekeeping 207
  and scout bees 60–1
  and spring breakthrough 31
  and summer bees 56
  and bee transfer of 134
  and travelling bees 143
  and wax moths 221
pollen baskets 131
pollen bundles 56, 131

INDEX

pollen combs 56
pollen crescent/circle 13
pollen dust 131
pollen supply 60, 208
pollination 134, 198
pollination apiary/beekeeping 205–8, 216
propolis
   and apitherapy 120
   and beekeepers 130
   and comb building 19–20, 127–9
   and comb cell rims 22, 26
   and combs 211
   and foragers 61
   and scent, odour, perfume 234
protein 31
pseudoscorpion 220–3
pupae 29–30, 45, 49, 186, 213
pupal case 126, 221
pupal cells 30, 39, 67, 116, 210
   and epigenetics 95
   and heater bees 45
   and learning capacity of bees 116–17
   and the Varroa mite 189–90
   and warmth 49
pupal stages 29, 39, 67, 95
purgative flights 36

queen 2, 22, 33, 40, 124, 157, 170, 213 *see also* pheromone, scent; royal jelly
   and beekeepers 179–80
   and comb cells 13, 20
   and death of 171, 175–6
   and drones 88–9
   and egg laying 95, 103, 153, 158, 171, 175
   and sexual maturity 102
   and nurse bees 39, 98, 176
   and offspring diversity 94
   and the queen cell 101
   and queen substance 99–100, 149
   and spring breakthrough 36
   and swarm time 160–1
queen cell bowls 101
queen cells 101, 124, 160, 170–1, 174, 176, 180
queen excluder 148–9, 151, 153, 173
queen nuptial flight 100, 102–3, 160, 171–2, 175–8, 186
queen replacement 100–1, 176–9
queen substance 99–100, 149 *see also* hive harmony; pheromone, scent
queenless colony 100, 176
queen's court 39

INDEX

rapeseed honey 133
Reaumur, Rene-Antoine Ferchault de 53
Reinhard, Judith 115–16
relative humidity 148, 207
removable frames 12, 15–16, 148
  see also frames
robber bees 186, 217, 195
royal jelly 40, 95, 101, 103, 120, 123–4, 160, 176–7, 180 *see also* nurse bees; nurse's milk; queen

scent, odour, perfume 59, 70, 80–1, 85–6, 109–11, 141, 165, 227
  and antennae 45
  and communication 22
  and dancers 140
  and disease in the hive 211
  and drones 88
  and foragers 79
  and guard bees 57
  and propolis 234
  and scout bees 61, 136
  and swarm time 161
scent trails 60–1, 83, 88, 164, 167
scenting posture 78–9, 161
scout bees *see also* bee dance; waggle dance
  and nectar 68, 135–8, 185
  and piping 165–6
  and scent, odour, perfume 61, 110, 136
  and sourcing supplies 60
  and swarming 162–4, 166–7
selfish genes 178–9, 230–1
separator 150
sexual maturity
  of drones 88
  of the queen 102
  of the Varroa mite 190–1
sexual reproduction 90–2
skeps 10 *see also* basket apiaries
smoke effect on bees 223–5
solitary bees 203
sperm 87–8, 90, 92–3, 100, 103
spring blossoming 52, 159, 185
spring breakthrough 36, 38, 39, 40
spring honey 205, 209
spring season 29, 37
starvation
  and the bee colony 153, 236
  and drones 90
  and the queen 104
stearin 11, 125
sting 107, 121–3, 176 *see also* apitoxin; bee venom; venom sac
  and guard bees 57–8
  and the queen 102, 171
sting bristles/setae 120, 122–3
sucrose 145

INDEX

sugar syrup 31
    and winter feeding 154–5
summer bees 38, 168, 213
    and foragers 61–2
    and honey-making bees 56–7
    and life cycle 51–2
super 14–17, 52, 128, 148, 173 *see also* feeder super; honey super; magazine hives
super mite 193
swarm capture basket 10, 169, 174
swarm cells 160, 174–5
swarm cluster 161–6, 169
swarm drive 159–60, 170, 173–5
swarm formation 167
swarm intelligence 161–2, 164–8
swarm mood 170
swarm period 160, 195
swarm prevention 173–4, 181, 206, 218
swarming 6, 9, 175, 218
    and bee vision 107
    and comb building 18, 222
    and drones 89–90
    and magazine hives 172
    and scout bees 162–4, 166–7
swarms 59, 102, 158, 171–3, 177
    see also afterswarms
    and the bee dance 26
    and bee recruits 165
    and beekeepers 10, 169, 172, 178–9, 218
    and the queenless colony 100
    and swarm intelligence 161–2, 164–8
    and swarm mood 170

teamwork 4, 29, 44, 67, 104, 161
thermal radiation 50–1
thermometers 33, 49
thermo-physical effects 225
travelling apiaries 143
travelling beekeepers 142
travelling bees 143
tree hollow hives, 7, 17, 218, 220–2, 225
    and propolis 128
    and relative humidity 148
    and scout bees 163
trophallaxis 43, 117–18

University of Wurzburg 114–15

Varroa infections 191, 203
Varroa mite 187–9, 190–2, 195–6, 201, 210, 212, 216, 220, 223
venom glands 57
venom sac 121–4
ventilator bees 95–7, 146

INDEX

vibration *see* comb vibration
visual sense *see* bee vision

waggle dance 23, 61, 67, 137, 140, 163 *see also* bee dance
    and 'dance communication' experiment 68, 70–1, 73–5
warmth 8, 18, 34–5, 42–9, 49–51, 213 *see also* brood areas; brood cells; brood nest; heater bees
    and piping scouts 166
    and the winter cluster 32–5, 41
wax *see* beeswax; wax plates
wax harvesting 10–11, 125–6
wax moths 221–2
wax plates 17–18, 21 *see also* beeswax; cell caps; cell-cap wax
Weippl, Theodor 65
wild bees 172, 203, 219, 220, 222
willow blossoms 131
willow trees/Salix 2, 37
winter bees 29, 213
    and brood nest temperature 95
    and last foraging flight 38
    and life expectancy 30
    and metamorphosis 37
    and the purgative flight 36
    and the queen 33
    and winter nourishment 31

winter cluster 32–3, 34, 35, 36, 41, 235
winter feeding 153–4
winter reserves 8, 31, 33, 37, 153, 224
winter solstice 31, 35, 232
worker bee cell 21, 38, 176–7, 189–90, 192, 195
worker bees 21, 37, 41, 65, 184
    and bee sleep 64
    and the brood nest 47
    and chromosome sets 91, 94–5
    and comb building 52
    as drone mothers 177
    and drones 87, 89–90
    and egg laying 178–80
    and foraging flights 67
    and honey collection 149
    and life cycle 63–4, 213
    and cocoons 38
    and the queen 99–102
    and 'sweating' wax plates 17–18

Zander dimensions 16
Zeidel forest 225
Zeidler method 220–2
Zhang, Shaowu 113